SPARTON
POLO CLUB
SUPERHETERODYNE
ZENITH
BROADCAST
METERS
SHORT WAVE
Sentinel
Sparton
KILOCYCLES
DETROLA
EGM
KADETTE
EKCO RADIO
MÜNCHEN
HAMBURG
BERLIN
BRESLAU
LEIPZIG
WIEN
PRAG
BUDAPEST
INNSBRUCK
DRESDEN
KÖNIGSBERG
FRANKFURT M.
GLEIWITZ
NÜRNBERG
LANGENBERG
SHORTWAVE
POLICE

DECORADIO

the most beautiful radios ever made

Peter Sheridan

Schiffer Publishing Ltd
4880 Lower Valley Road • Atglen, PA 19310

This book is dedicated to my father, Ernst Schartenberg (1905–1957), who took the name Ernest Sheridan to join the English Army in 1940. His turbulent life treads a pathway through the years and countries explored in this book. He lived in Germany in the 1930s, England in the 1940s, and Australia in the 1950s. He is my shadow listener throughout and I can picture him with these radios, enjoying his music, news, and sports while smoking his ever-present cigarette.

Library of Congress Control Number: 2014938005

Front cover: Saprtan 500C "Cloissone," USA 1939, designed by Walter Dorwin Teague
Title page: Ekco AC97, UK 1936, designed by Jesse Collins

Photography, text, and design by Peter Sheridan
Typset in: Futura Std

ISBN: 978-0-7643-4605-7
Printed in China

Published by Schiffer Publishing, Ltd.
4880 Lower Valley Road
Atglen, PA 19310
Phone: (610) 593-1777; Fax: (610) 593-2002
E-mail: Info@schifferbooks.com

CONTENTS

FADA

FOREWORD

Peter Sheridan is known for his wonderful radio collection and his love of Art Deco. He is a respected historian and author in the Art Deco field and is renowned for his lectures, articles, and beautiful photographs. This book highlights both his photography and his scholarship as he combines the worlds of radio, Art Deco, and industrial design with a story that has never before been told. The radios alone are gorgeous in their streamlined beauty, but the number of famous designers involved and the clear extension of the Streamlined style throughout the radio industry worldwide during the 1930s and 1940s is an eye-opener. For lovers of radio, Art Deco, or industrial design, this book will be an enduring and beautiful reference.

Kathryn Hausman
President Emeritus
Art Deco Society of New York

In 2008, Peter produced *Radio Days,* a seminal work that celebrated the great range of Bakelite radios in Australia. Thankfully, he has now extended his work to look at Art Deco radios on a global scale. Naturally, much of the content is concerned with the USA, but areas such as Europe, England, and Australia are also covered extensively. The new work again pays attention to detail in the stories of the radios and is enhanced by excellent photography.

An exciting aspect of the work is the recounting of the role played in radio design by leading American and UK industrial designers, such as Loewy, Bel Geddes, Teague, Coates, and Chermayeff. This aspect of their work has long been understated, but is now fully revealed. In March 2013, the audience at the 12th World Congress on Art Deco in Havana sat entranced as Peter explored the relationships between industrial design, materials, manufacturing, and Art Deco styling that accompanied the massive global adoption of radio. Undoubtedly, radio played a major role in the spread of rapid communications across the world, but at last the role of the radio as part of the history of industrial design and Streamline styling is assuming an appropriate level of importance. This book will play a major part in this process and Peter is to be congratulated for his vision and extraordinary efforts to bring this story to the world.

Robin Grow
President
Art Deco and Modernism Society
Australia

ACKNOWLEDGMENTS

Hopefully, *DecoRadio* is a timely sequel to five important books from the 1980s and 1990s: John Sideli's *Classic Plastic Radios of the 30s and 40s;* Philip Collins' *Golden Age of Radio* and *Radio Redux;* Robert Hawes' and Gad Sassower's *Radio Art* and *Bakelite Radios*. Each of these books is an enduring and vital resource, reflecting the passion and expertise of the authors. My first Catalin and Bakelite radios came from Gad Sassower's shop "Decodence" in Camden Passage, London in 1998—a green American Emerson AU 190 (p. 183) and a white marbleized Australian AWA Radiolette (p. 144). From the moment I saw these radios I was besotted, and fifteen years later the infatuation remains.

As in my 2008 book on Australian Bakelite radios, *Radio Days*, the majority of radios in *DecoRadio* are from my collection, but it was clear early on that there were treasures I could not acquire and were necessary to ensure this book truly and completely illustrated my thesis. So I reached out to dealers and collectors, some of whom I knew and many whom I simply approached via their websites. With no exceptions, everyone was supportive and generous, many providing much more material than I asked for or expected. Everyone who was contacted exhibited an extraordinary willingness to add to the pool of knowledge and data that underpins this book. These old and new friends and colleagues supplied advertisements, patent diagrams, and ephemera as well as radios. They are listed below, and I gratefully acknowledge that this book owes much to the support and encouragement of these enthusiastic and knowledgeable radiophiles.

Hugh and Jane Hunt kindly gave me access to their wonderful collection, to photograph those radios I didn't have. I thank them for their hospitality and kindness. Thanks to Robin Grow, President of the vibrant Art Deco and Modernism Society of Australia, who has been a long-time friend and supporter. He has provided opportunities for me to publish monographs on radios and other Art Deco subjects in the world standard *Spirit of Progress* journal. *DecoRadio*, while being about radios, maintains an Art Deco theme and ADMS has been very supportive of this fundamental view of radio as an Art Deco icon.

The following people all provided radios, photographs, advertisements, information, and assistance for which I am truly grateful.

Hugh & Jane Hunt, USA
Steve Turner (Catalinradio.com), USA
Mike Schultz, USA
Bill Whelan (oldradiocrazy.com), USA
Gaelle Reboul (decophobia.com), USA
Mike Kinerk and Dennis Wilhelm, USA
Steve Craig, USA
Ron Ramirez (philcoradio.com), USA
Michael Raanan, Australia
Ian Revelman, Australia
Hans Hoogvliet, Australia
Mike Cooper, Australia
Steve Milliken, Australia
Ritchie Singer, Australia
Scott Beveridge, Australia
Robin Grow, Australia
Carl Glover, UK
Philippe & Chris, Belgium
Johan Catteau, Belgium
Roland De Smet, Belgium
Emiel Franck, Belgium
Freddy Crabbé, Belgium
Rudi Sillen, Belgium
Pierre Pechon, France

In the end, nothing would have been achieved without a wonderful wife who gave me the time and space to concentrate on this second book on radios, and also encouraged me to build our collection of the best possible examples that reflected a world view of radio. Jan Hatch is a gifted stylist and curator, being able to place 250 radios discretely around our apartment such that they do not overwhelm the décor or our lives. As I have said on many occasions, Jan can make an obsession look like a noble quest. Often those who support are never given adequate kudos, so here in perpetuity I say to Jan, *"thank you."*

NOTES TO THE READER

The photographs in *DecoRadio* can be used simply as a reference for collectors, although there has been no attempt to provide a complete catalog of radios from the period. Radios have been chosen that best represent the global expression of the Deco style and particularly those created by important industrial designers. Not every radio is a perfect, original example and some may have repairs and modifications that I have either missed or ignored that will be glaringly obvious to the specialist collector or dealer. In many instances, rarity and availability trumps perfection. However, I have tried to avoid fakes (or those with a questionable provenance), cabinets with major alterations, and those radios exhibiting colors, schema, and finishes that were never available at the time of original sale.

The Radio Gallery section is presented chronologically by decade. As a preliminary to each decade, I have provided a series of thumbnails of the radios of the period listed in sequence and by country. As well, at the end of the book is a Gallery Index sorted by country, maker, model, and year. This will help readers to quickly gauge the date of a radio and roughly where in the book it is situated.

One of the great confusions in books on radios is judging the actual size of the radio from the image on the page. Rather than provide detailed information about dimensions, the thumbnails give a quick visual approximation of size. The text chapters can be read independently of the Gallery section, but I have inserted small photos in the margins for visual context and to help tell the story.

Model names: Model names that were associated with the original release of the radio are placed in quotation marks; for example, Emerson "Patriot." Where a later name has been in general use by the public or collectors, I have used enclosing brackets; for example, Emerson (Mae West).

The lack of footnotes and precise references is a calculated attempt to make *DecoRadio* more accessible to a wide audience. The subject matter is probably original enough to be worthy of a formal thesis and some academics may underrate the content because it is not supported by an evidence-based approach. However, as a senior health-professional with a lifetime of teaching, editing journals, and writing articles in many fields, I chose, in this case, to dispense with some of the trappings and veneers of academe to simply present the material in a logical and, hopefully, interesting manner. In fact, I have a cynical view that scientific research publication has become an end in itself, a vacuous measure of poor methodology, short time-frames, and inadequate sampling. Reliance on such a questionable base of evidence may actually disguise a lack of critical thinking. The reading list will attest to the breadth of my enquiries. Trolling the Internet brought up interesting commentary, occasional original information, and often confirmatory details. This has to be balanced against the cut-and-paste nature of the web, where much data assumes the mantle of veracity because of repeated exposure on different sites.

I have endeavored to provide precise dates for manufacture or release of radios in *DecoRadio,* but complete accuracy is difficult. Some models were sold over a number of years and a few sold both before and after World War II. The Radio Museum website is a particularly reputable and valuable source, as it checks the date of manufacture by reference to accessory data such as the valves used in the radio.

Attempting a display of radios from all over the world almost assuredly will mean that some special and beautiful examples have been omitted. There is probably a second book of those radios I missed and if many more beautiful radios come to light because of this book, then that is a bonus. There will be a website at *decoradio.com,* and perhaps this is the place for additional Art Deco radios to be displayed as producing a hard copy update of *DecoRadio* will probably be prohibitive.

Almost all photos were taken with a Nikon D800 DSLR with a 24–70mm 1:2.8 Nikkor Lens and bounce flash via a Nikon SB 800 speedlight with diffuser. All images were taken with RAW files and layered in Photoshop CS6. Colors of radios are "as shot" with an express avoidance of enhancement and alteration.

The image of *"Phenolica—The Goddess of Plastic radios"* needs a little explanation. The original image was an exit sign in a defunct Deco cinema in Montreal. Courtesy of Photoshop, the round radio replaced a silver ball in the lady's hands and the word "radio" at the base covers the original exit sign. Phenolica may remain unresponsive to all entreaties but her earthly representative can be contacted at peter@decoradio.com.

Phenolica—Goddess of Plastic Radios

CHAPTER 1
INTRODUCTION

This book has a number of objectives and themes around the notion of "the most beautiful radios ever made." Rather than detailing the technical development of this medium, this is a different view of the burgeoning world of radio as it became the entertainment and information center of every household. The focus is on design with radio as a leading player in the evolution of a new style movement. Accompanying the story is a photographic display of extraordinary radios that will be a delight for collectors and a surprise to everyone else.

As a devotee of Art Deco, I have always seen these pieces as quintessential representations of the style, yet almost completely ignored in any description or history of the Deco genre. This is perhaps understandable given that these radios were generally mass-produced with new materials and often marketed as inexpensive tabletop alternatives to the more prestigious wooden consoles. But with evidence that the most famous industrial designers in America and England were at the forefront of this new wave of modern Streamlined radio cabinets, it now seems untenable to discount their significance. This book champions the often maligned wireless and seeks to elevate these domestic appliances of the 1930s and 1940s to the status of important icons in Art Deco history and design.

Bakelite was the first synthetic plastic and from this began a resin revolution that, within the space of 100 years, has changed the way we live. In many ways, the Bakelite and Catalin radios of the 1930s were the climax of this first iteration of plastics. This period saw radios reflect the change in the manufacture of household and decorative items from organic materials to synthetics, from the work of the single artisan to mass production and from a range of browns to a rainbow of colors.

The early 1930s, in both America and England, saw designers bring their talents into the industrial arena. More recognized for other achievements, both at the time and now, they nevertheless introduced elegant Streamlined designs for radio with colored synthetic resins, glass, and metal, as well as the traditional wood. In many cases, the plastic radios were the first mass-produced Deco style colored plastic objects in people's homes. Not only did the designers influence the public to purchase these novel shapes and variegated shades, but the innovation and functionality created a benchmark and a trend for all other manufacturers of radios in the world to follow.

Radio was an immediate success as a vehicle for communication, being quickly and broadly taken up in the Western world. While this might be put down to the capability of this new tool and the varied broadcast content that appealed to all members of the family, there are other factors that needed to be in place for the medium to flourish. At least in part, the extensive distribution of radio required the confluence of a number of diverse elements. Radio could not be a household product until it moved from being a novelty to a necessity. As a priority, this required an integrated and user-friendly apparatus of an appropriate size. New materials, new designs, and mass production allowed a cost-effective home appliance to be mass marketed. The advent of the mantel radio in 1930 gave a huge impetus to the spread of radio, not only allowing multiple radios in the home and workplace, but changing the listener from the family to the individual. A choice of colors allowed for targeted marketing directed at the woman of the house to place radios in rooms other than the lounge. Finally, the spread of domestic electrification also underpinned the expansion of radio sales and usage.

It is remarkable that during the inter-war period and the Depression, when unemployment and poverty were common, the radio flourished. The uptake of radio within the Western world was swift and complete and it is remarkable how much people were prepared to spend on a radio in times of financial difficulty. This says much for what it represented as a new technology and its impact on the lives of all members of the community.

Today many of the Bakelite and Catalin radios have moved from lowly collectibles to respected antiques, but this should not be seen out of context. There are a majority of plastic radios that have no special redeeming features and even more wood cabinets that are merely unremarkable packages for a radio chassis. Wood can be attractive, particularly with the use of veneers, and a number of fine examples with Deco styling are included in *DecoRadio*. But, in general, wood cabinets were restricted by the material, production methods, and lack of color variations. On the other hand, the once vulgar plastic radio is now accorded prestige status by the auction houses with high prices achieved for sought after models and colors.

The increasing appreciation of the Bakelite and Catalin Deco radios of the 1930s may be that this singularity is now being seen as an extraordinary and exciting prelude to our current fascination with new communication technology, our expectation that everything is available to us, and of course the ubiquitous application of plastics in our lives. We are attracted to these radios because they are chic today and yet they are from the past. We see them shining in the houses of yesteryear calling out that, unbeknown to their owners, *"we are the future!"*

To many radio enthusiasts, this book will be an incomplete view of radio history. Some will see it as superficial, the radio cabinet being merely an accessory to the main role of communication. But, as some of us do with cars, we fall in love with the exterior styling and even the color without having to know what sort of engine is under the hood and how it works. That is not to say that the technology is unimportant and its development is of course a worthy topic. Similarly, the content and the characters that inhabited the radio ether over the years deserve to be chronicled. But these subjects have many supporters and their cause needs little additional scholarship. This book focuses on and serves as an advocate for the radio cabinet, particularly those formed from plastic, making a case for its importance in the story of radio.

As an Australian, I know instinctively that "Aussie" radios are only a part of the whole story. This is a benefit of our cultural cringe, the awareness that what we love and hold dear may be only part of a much bigger picture. With this sort of premise, the grand saga is retold without national prejudice and much of the oft-repeated "we were first" claims are recast in a more objective and global overview. Of course, I have made some assumptions and inferences, which may be disputed, but I ask some indulgence on the basis that my prevailing prejudice is simply that of a lover wishing to portray and describe my international paramours, showing off their best features. I accept that these beauties were fragile, noisy, often unresponsive, overheated, and their lustre ultimately faded. That so many retain their essence after some eighty years and are more and more appreciated is a measure of their intrinsic significance and inherent elegance.

The radios presented are (in the main) a personal choice from my collection. I have photographed those that I believe will give you, the reader, a chance to sample the best the world produced during that short twenty-five year period of the 1930s to the early 1950s. Some are quirky, some help tell the story, some are truly unique, but all share that Deco style, which has a timeless elegance, and are thus, in my eyes, and I hope yours, truly beautiful.

CHAPTER 2
SETTING THE SCENE

The late nineteenth and early twentieth centuries ushered in revolutions that changed people's relationship with their physical environment and shattered the insularity of family and community. The engine gave people (either singly or *en masse*) the ability to travel by car, train, ship, and plane across vast distances in a manner and speed unimaginable a few decades earlier. Communication by telegraph, radio, and film changed the manner of societal interaction, giving all members of the family access to entertainment and information separate from the normal pathways of family, school, and church. New materials, such as plastics, aluminum, and stainless steel, replaced traditional organic materials and were utilized in the mass-production of all sorts of consumer-oriented products.

In the mid-1920s, radio was still a novelty. By the late 1920s, radio was clearly becoming more commonplace, and, by the mid-1930s, a ubiquitous, user-friendly home appliance. The radio of the 1930s and 1940s could be described as a metal chassis with valves (i.e., tubes) and a speaker surrounded by a container of wood or plastic in various configurations that transmitted information and entertainment from local and remote sources to a domestic audience. The radio was the modern technology that brought the world into the home. It was a selective convenience used when desired to provide programs ranging from news to sport, music, drama, and education, all interspersed with advertising and occasional political indoctrination. In general, it was an extraordinary vehicle for broadening the lives of all members of the family.

Radio brought about a third wave of the democratization of information (the advent of speech and the printed word being first and second). Newspapers, magazines, and books were the predominant communication pathway of the nineteenth century, but in truth, literacy rates universally were not high. Radio enfranchised the least educated with information they could assimilate. It was actually cheaper to provide radio waves than newspapers in remote areas and radio more easily reached those with learning difficulties than newspapers. In America, in the late 1930s, the radio became the main information provider for low education groups and rural listeners. By 1937, seventy percent of the American public reportedly depended on the radio for their daily news and a majority considered radio a credible media. In the 1940 presidential campaign in the USA, fifty-two percent of a cross section of American voters got their major information from the radio and thirty-two percent from newspapers. Radio is still regarded as showing the fastest and most complete uptake of all the new technologies of the twentieth century, which include the telephone, TV, video/DVD, and the Internet.

In the main, radios presented as bland and boxy, often mimicking the excesses of Victorian, Edwardian, or even Gothic fashion. With cabinets mainly of wood or wood-colored plastic, they were produced in volume for economy rather than quality or style. Differences in technology were secreted in the valves and circuitry, improving reception and amplification, and this was the main feature promoted in sales advertising. Its utility as a communication device was paramount, a unique device of modern wizardry with a knob on the left for on/off and volume, another on the right for changing the radio station. Occasionally, a third knob was available for finding long, medium, and short wave stations, which expanded the range of available broadcasts from provincial to global.

The radio was often a considerable piece of furniture in the lounge or main room of the house. By the mid-1930s, the mantel model was a competitor as well as a sibling, a smaller and more portable appliance suitable for any room in the home. For any invention to succeed, the timing must be right. Radio could not have spread across continents and into the vast majority of homes in less than two decades without the additional infrastructure of mains electricity. This state resource, at first in the cities and then slowly radiating out to rural areas, gave radio expanded versatility—the impetus to be ubiquitous—in every house and even in every room. As more and more homes and businesses acquired electricity, and when every room in the house and every place of employment became a suitable listening post, the mantel radio became even more popular than the floor standing console. In the 1930s and 1940s, most houses had one phone and perhaps one car, but a large proportion of homes had more than one radio.

All of this happened in the context of global social upheaval involving financial catastrophe and warring states. Although in hindsight we see the 1920s and 1930s as an interwar period, the prospect of World War II was not a pressing issue in the early 1930s, although the effects of the stock market crash were significant. New technology and widespread industrialization in the early decades of the twentieth century engendered significant wealth to a few, but in most countries the workers struggled with poverty and a modest standard of living. The Depression added to the pool of the poor, and the early 1930s appear a difficult time for manufacturers to encourage consumers to make purchases that were not associated with the daily necessities of living. Demand for radios was growing rapidly in the late 1920s (7.5M units in 1928 to 11.8M units in 1929). However, the effects of the Depression, coupled with the advent of mains electricity, more complex circuitry, patent issues, and new shapes and styles of radio caused the number of radio manufacturers to contract by ninety percent in the next four years.

However, we need to see the first half of the twentieth century as setting the core parameters of modern living, a huge change in the society's expectations based in a major part on the impact of technology and science. One singular adjustment was the veneration of the individual. Combined with the breaking of rigid class strictures after World War I, the 1920s and 1930s in many western countries was a period of great optimism regarding personal rights, advancement, and potential. The rise of communism and fascism further served to polarize views and approaches to individual rights. It is no surprise that the United States, England, Australia, and Canada are environments where radio design was most expansive and uptake most complete. In Germany, the use of radio was hijacked for the distribution of political propaganda, and there is a noticeable contraction of design options and style innovations.

In 1930, people embraced the prospect that the machine age and speed would improve their lives, and that divesting themselves of traditional decorative strictures was part of the new uniform of modernity. In spite of the Depression, and perhaps even because of it, people looked to the future for encouragement. New communication media, such as the telephone, the radio and movies, all provided emancipation (albeit sometimes brief and virtual). Transport, both private and public, offered mobility and opportunity. New technology and mass production offered the public access to many items previously either unknown or only available to the rich. Not everyone could afford a car, but most could buy a radio, even if they paid by lay-by.

Common wood and Bakelite radios of the 1930s and 1940s

It might at first glance be hard to justify why such ordinary objects as radios should become iconic in the Art Deco pantheon. For this to be an accurate appraisal we have to ignore ninety-five percent of the radios produced at that time and concentrate on a small subset of valve radios that appeared in 1930, found full flowering between 1933 and 1940, and had a short resurgence after World War II. By 1950, other styles and technologies became paramount. In some countries, the timing took a shift to the right, indicating upheaval and retarded development of capacity and progress, thus delaying expression until the mid-1950s. The radios presented from Russia and Czechoslovakia in *DecoRadio* are good examples of deferred modernization.

Most of the motor vehicles manufactured since 1900 have no Art Deco styling; however, we can focus on certain makes and models from the USA and Europe in the 1930s and 1940s—the American Cord and the Spanish Hispano-Suiza quickly come to mind—and feel comfortable assigning them as archetypal expressions of the Art Deco movement.

In the same way, we need to concentrate our attention on a small, somewhat lost group of radios that have a clear integral coherence in their evolution and show classic features of Art Deco influence. They only represented a minuscule proportion of global radio production and sales, but by virtue of their aesthetic attributes they influenced all radio cabinetry to some extent.

Items like cars have not dramatically altered in concept or reality since their original development and the evolution of today's models can clearly be recognized as a linear development. Yet radios have undergone a radical metamorphosis of size and technology, leaving the valve radio a nostalgic relic of a previous era. The radios we seek to present as demonstration of genuine and primary Art Deco design sold in their thousands, but against a backdrop of millions of sales. They were generally not highly priced, were made of cheap materials, and often had inferior performance features. Their subsequent and continued dismissal by radio enthusiasts is understandable.

Quite a few of the radios in *DecoRadio* are so rare as to be unique one-off examples produced as samples or in very short production runs. Many did not sell well, being too outlandish for the consumer's tastes and inconsistent with home décor of the day. For instance, Ekco's colored radios in the mid-1930s simply were unacceptable to the British populace and this preference for brown did not change until the early 1950s.

It is notable that these radios were perhaps the first colored Art Deco items in people's homes in the 1930s against a general backdrop of post-Victorian, Edwardian, and Art Nouveau domestic interiors. Art Deco architecture was not a universal phenomenon, often being limited to large public buildings in major cities. At the domestic level, even where external features displayed modern Streamlined elements, interior treatments often reflected a mixture of earlier styles.

Thus, most radios (like cars) are bland and unprepossessing. Unlike the special Deco style vehicles that have been identified and iconized, the unusual Deco radios of the 1930s and 1940s have rarely been seen in public and are generally only known to collectors, who often have little interest in the Art Deco heritage. Yet, when placed in proximity to each other, these radio gems clearly reflect a coherent narrative and development that is Art Deco inspired and perhaps even inspiring of Art Deco.

CHAPTER 3

THE MOST BEAUTIFUL RADIOS EVER MADE

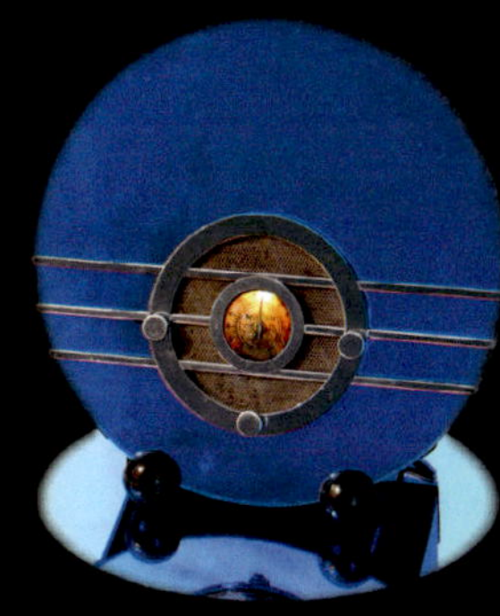

There are eras in technology, manufacturing, and design that are respected and recognized for their place in the evolution of our society. The advent of the telephone, the car, and the skyscraper are examples that come to mind. Most of these revolutions have been well documented and the subject of many books, documentaries, and critical analyses. We clearly recognize the advent of radio as an epochal event and its uptake into the community both unparalleled and pervasive. There is much written about the development of the technology of radio and, from a social aspect, plenty about the characters and programs that caused families to gather round the radio every day. But there is a part of the story of this extraordinary communication device that has been lost in its translation to the commonplace.

The years from 1930 to 1940 were a pocket of opportunity wedged between the deprivations of the Great Depression and upheavals that presaged the advent of World War II. The common image of radio before its miniaturization in the 1960s is that of a drab piece of wooden furniture or an unremarkable brown plastic box. But there is a period of innovation and glamor that seems forgotten, except by a small group of collectors who hold the last specimens of this important lineage. Fuelled by the genius of industrial designers, the advent of new materials, better manufacturing processes, plus marketing and consumer changes, there was a golden age of radio that today is largely overlooked, both by the public and by those who record important changes in our society.

In recent years we have revisited the story of the dinosaurs, those lumbering pea-brained monsters whose name is synonymous with all that is not only outmoded and impotent, but extinct. We now recognize that they were not just a diverse group of reptiles that suddenly disappeared, but that they were indeed the forerunners of today's birds. Just as surprising is the revelation that in all their reptilian diversity over 60 million years, they were not all dun colored, but many were brightly and variously colored.

In the 1960s, as transistors and miniaturization transformed radio into a vehicle that reflected its core function and eliminated any need for an aesthetic external case—the valve radio truly became a metaphoric dinosaur. But if we look back to the decade from 1930, we see a remarkable evolutionary transition took place which extends and parallels the dinosaur analogy in terms of diversity of shape and color. Also, the colored designer radios were relatively insignificant in terms of numbers and impact. It is only in retrospect we see their importance and can elevate their status to iconic.

Radio's main function is broadcasting, yet it served also as a piece of furniture and a domestic appliance. There are few technological advances that had so great an impact, and yet were so varied in presentation. The telephone had a single communication function and remained limited by a core design and almost universal color scheme. The car, on the other hand, is a worthy contender for title of most iterative technology—it is a mode of transport displaying a myriad of models, functions, and eventually many colors—but it does utilize a limited palette of external structural materials.

We rarely appreciate that the radio cabinet came in wood, plastic, metals, and even mirrors, or that the cabinet design ranged from floor standing consoles to tabletop models and portables. They varied in shape from vertical cathedral and tombstone shapes to architectural styles and boxes of varying configurations. Some were even produced in circular form. In design they reflected Victorian, Edwardian and, most importantly, Machine-age, streamlined Art Deco styles. Valve radios ranged from two inches (50mm) to four feet (1200mm) high. Although for some wooden radios there were elegant mixes of veneers it is only with Bakelite and Catalin that the colors reflected the whole rainbow spectrum. One highpoint of coloration was the extraordinary "mix and match" approach with the color of the grill, knobs and handles contrasting that of the body of the radio. This was truly an American approach, simple yet effective, colorful and diverse. This styling seems so current that it is hard to believe these radios are over 75 years old.

The majority of radios sold in the 1930s and 1940s were wood or brown Bakelite. This seemed to be based on a social rather than a cost decision, reflecting standard home décor, ideas of furnishing and color schemes. Console radios, particularly were pieces of furniture, with the radio often intentionally concealed or secondary to the cabinet itself.

The beautifully designed and colored small radios were then and are now a rarity. They were a subset of a subset—an alternative choice usually at the cheaper end of a company's radio range. For example, in the brochure of Emerson Company's range of radios for 1939 (see right) there are thirty-five models listed. Of these, twenty-seven are wood, seven Bakelite, and one Catalin. Of the twenty-seven wood, six are consoles (ranging from $69–$169) and the remainder table models. One of these is the famous Emerson (Mae West) at $39.95. Of the seven Bakelite table models, four are brown and three ivory. The only Catalin radio is the AU-190 Cathedral in the "standard" line at $24.95 and offered in "six radiant colors." As this model was released in 1937 and there is no mention of other Emerson Catalins released in 1937 and 1938 (BT245 Cathedral, BN258 "Big Miracle," and AX235 "Little Miracle"), one can only conjecture that they were not sold in large quantities. Also, the position at the bottom of the "standard" line says much for its position in the 1939 radio hierarchy.

Similarly, Motorola lists thirty-four radios in its 1941 range (see right). There are nineteen wood, ten Bakelite, four faux-leather, and one Catalin. Of the nineteen wood, four are consoles and fifteen table models. Of the ten Bakelite, all are table models with six brown, three ivory, and one brown/ivory. The four faux leather are all battery portables. The one Catalin is the very special 50XC "Circle Grill" model offered in four "bright plastic" color combinations. The 50XC was released the year before in 1940 and there is no mention of Models 52 (1939) and 51X16 (1941). This does show an increase in Bakelite at the expense of wood for table models in two years, but all consoles continue in wood and colored radios contribute only a single entry in both these two examples. It is probable that there were more radios available than listed in these brochures, but it is hard not to conclude that colored radios were a small part of a very large market and not necessarily high on the promotional strategies for radio manufacturers.

Over time, colored radios were generally ignored by those who enjoyed tinkering with valve radios. They were

Emerson AU-190

Emerson BD-197

Motorola 50-XCI

EMERSON *De Luxe* LINE

Model BF-207 (Ivory) $29.95
With "MIRACLE TONE CHAMBER"
6 Tubes (including ballast tube) AC-DC Superheterodyne. American, Foreign, Police. 6½-Inch Dynamic Speaker. Automatic Volume Control. Tone Control. Gemloid Dial. Beam Power Tube. Emerson Foreign Tuning Principle. Beautiful ivory bakelite cabinet.

Model AA-207 (Ivory) $34.95
16 to 2,000 Meters. Otherwise same as model BF-207.

AU-213 $29.95 "MIRACLE TONE CHAMBER"
5 Tubes AC-DC Superheterodyne. American Broadcasts, All Police, Amateur, Aeroplane Stations. Electro Dynamic Speaker. Automatic Vol. Control. Tone Control. Beam Power Tube. Built-in Antenna. Staybent hand-rubbed walnut cabinet.

Model BF-204 $29.95
with "MIRACLE TONE CHAMBER"
6 Tubes (incl. ballast tube) AC-DC Superheterodyne. American, Foreign, Police. 6½" Dynamic Speaker. Automatic Vol. Control, Tone Control, Gemloid Dial. Beam Power Tube. Staybent hand-rubbed walnut cabinet.

Model AA-204 $34.95
16 to 2,000 Meters. Otherwise same as model BF-204.

Model BJ-214 $29.95
with "MIRACLE TONE CHAMBER"
6 Tubes (incl. ballast tube) AC-DC Superheterodyne. American Broadcasts, All Police Calls, Amateur, Aeroplane Stations. 6½" Dynamic Speaker. Automatic Vol. Control. Gemloid Dial. Beam Power Tube. Built-in Antenna. Staybent hand-rubbed walnut cabinet.

Model BL-214 $29.95
5 Tubes AC Superheterodyne. Otherwise as model BJ-214.

Model BD-197 $39.95
"MIRACLE DIAL"
6 Tubes (incl. ballast tube) AC-DC Superheterodyne. American, Foreign, Police. 6½" Dynamic Speaker. Automatic Vol. Control. Tone Control. Beam Power Tube. Band Indicator. Staybent hand-rubbed walnut cabinet.

Model BF-169 $39.95
With "MIRACLE TONE CHAMBER"
6 Tubes (including ballast tube) AC-DC Superheterodyne. American, Foreign, Police. 6½-Inch Dynamic Speaker. Automatic Volume Control. Tone Control. Gemloid Dial. Beam Power Tube. Cabinet of hand-rubbed butt walnut and Oriental wood of Staybent Construction.

Model BW-231 $39.95
with "MIRACLE TONE CHAMBER"
With "MIRACLE DIAL"
6-Tube AC Superheterodyne. American, Foreign, Police. 6½-Inch Dynamic Speaker. Automatic Volume Control. Tone Control. Dynamic-Coupled Power Output. 5 Watts Output. Band Indicator. Hand-rubbed walnut cabinet of Staybent Construction.

Model BQ-228 $49.95
With "MIRACLE TONE CHAMBER"
"Miracle Dial" "Miracle Instamatic Tuning"
6-Tube AC Superheterodyne. American, Foreign, Police. 6½-Inch Dynamic Speaker. Automatic Volume Control. Tone Control. Dynamic-Coupled Power Output. 5 Watts Output. Band Indicator. Hand-rubbed walnut cabinet of Staybent Construction.

Model BU-229 $59.95
With "MIRACLE TONE CHAMBER"
"MIRACLE DIAL"
and "MIRACLE INSTAMATIC TUNING"
7-Tube AC Superheterodyne. American, Foreign, Police. Electron Ray Tuning. 6½-Inch Dynamic Speaker. Automatic Volume Control. Tone Control. Dynamic-Coupled Power Output. 5 Watts Output. Band Indicator. Hand-rubbed striped walnut cabinet of Staybent Construction.

EMERSON *Standard* LINE

Model BM-206 (Walnut) $9.95
With "MIRACLE TONE CHAMBER"
5 Tubes (incl. ballast tube) AC-DC, American Broadcasts, Police Calls. Electro Dynamic Speaker. Automatic Overload Control. Gemloid Dial. Beam Power Tube. Built-in Antenna. Walnut bakelite cabinet.

Model BM-206 (Ivory) $12.95

Model BM-215 $14.95
With "MIRACLE TONE CHAMBER"
5 Tubes (incl. ballast tube) AC-DC American Broadcasts, Police Calls. Electro Dynamic Speaker. Automatic Overload Control. Gemloid Dial. Beam Power Tube. Built-in Antenna. Walnut cabinet.

Model BB-209 $19.95
With "MIRACLE TONE CHAMBER"
and "MIRACLE INSTAMATIC TUNING"
5 Tubes (incl. ballast tube) AC-DC American Broadcasts, Police Calls. Electro Dynamic Speaker. Automatic Overload Control. Gemloid Dial. Beam Power Tube. Built-in Antenna. Hand-rubbed striped walnut cabinet of Staybent Construction.

Model BE-198 (Walnut) $19.95
With "MIRACLE TONE CHAMBER"
and "MIRACLE INSTAMATIC TUNING"
5 Tubes AC Superheterodyne. American Broadcasts, Police Calls. 6½" Dynamic Speaker. Automatic Volume Control. 3 Watts Output. Walnut bakelite cabinet.

Model BE-198 (Ivory) $24.95
With "MIRACLE TONE CHAMBER"
and "MIRACLE INSTAMATIC TUNING"
5 Tubes AC Superheterodyne. American Broadcasts, Police Calls. 6½" Dynamic Speaker. Automatic Vol. Control. 3 Watts Output. Ivory bakelite cabinet.

Model BJ-200 (Walnut) $19.95
With "MIRACLE TONE CHAMBER"
6 Tubes (including ballast tube) AC-DC Superheterodyne. American Broadcasts, All Police Calls, Amateur and Aeroplane Stations. 6½-Inch Dynamic Speaker. Automatic Volume Control. Gemloid Dial. Beam Power Tube. Built-in Antenna. Walnut bakelite cabinet.

Model BL-200 (Walnut) $19.95
5-Tube AC Superheterodyne. Otherwise as model BJ-200.

Model BJ-200 (Ivory) $24.95
With "MIRACLE TONE CHAMBER"
6 Tubes (including ballast tube) AC-DC Superheterodyne. American Broadcasts, All Police Calls, Amateur and Aeroplane Stations. 6½-Inch Dynamic Speaker. Automatic Volume Control. Gemloid Dial. Beam Power Tube. Built-in Antenna. Ivory bakelite cabinet.

Model BL-200 (Ivory) $24.95
5-Tube AC Superheterodyne. Otherwise as model BJ-200.

Model BJ-210 $24.95
With "MIRACLE TONE CHAMBER"
6 Tubes (including ballast tube) AC-DC Superheterodyne. American Broadcasts, All Police, Amateur and Aeroplane Stations. 6½" Dynamic Speaker. Automatic Vol. Control. Gemloid Dial. Beam Power Tube. Built-in Antenna. Staybent hand-rubbed walnut cabinet.

Model BL-210 $24.95
5-Tube AC Superheterodyne. Otherwise as model BJ-210.

Model BF-207 (Walnut) $24.95
With "MIRACLE TONE CHAMBER"
6 Tubes (including ballast tube) AC-DC Superheterodyne. American, Foreign, Police. 6½" Dynamic Speaker. Automatic Vol. Control. Tone Control. Gemloid Dial. Beam Power Tube. Walnut bakelite cabinet.

Model AA-207 (Walnut) $29.95
16 to 2,000 Meters. Otherwise as model BF-207.

AU-190 $24.95
6 Radiant Colors: 5-Tube AC-DC Superheterodyne. American Broadcasts, All Police Bands, Amateur, Aeroplane Stations. Electro Dynamic Speaker. Automatic Volume Control. Tone Control. Gemloid Dial. Beam Power Tube. Built-in Antenna. Plastic cabinet.

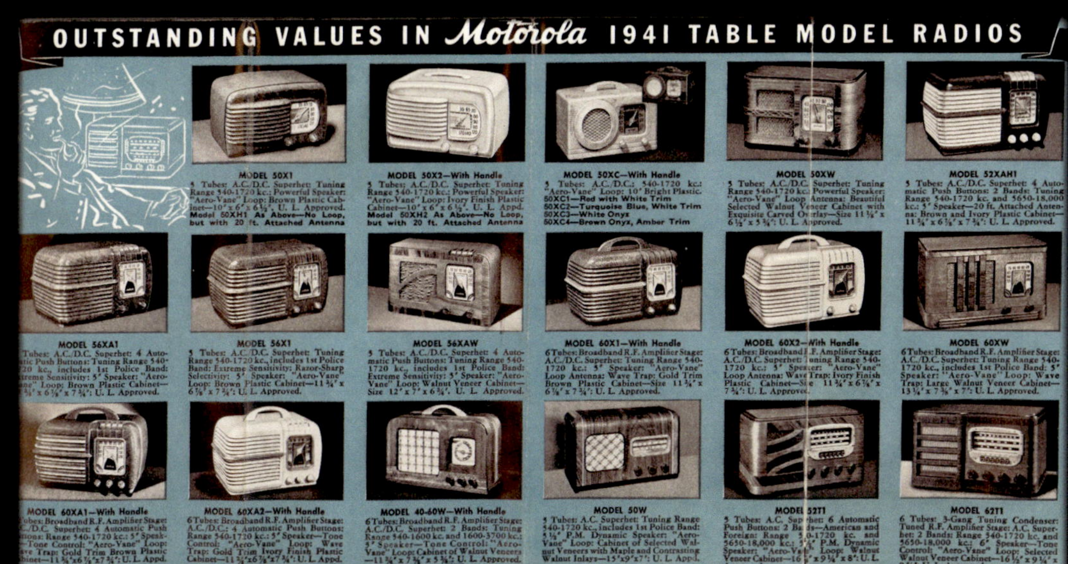

OUTSTANDING VALUES IN *Motorola* 1941 TABLE MODEL RADIOS

MODEL 50X1
5 Tubes: A.C./D.C. Superhet: Tuning Range 540-1720 kc.: Powerful Speaker: "Aero-Vane" Loop: Brown Plastic Cabinet—10″ x 6″ x 6½″. U. L. Approved.
Model 50XH1 As Above—No Loop, but with 20 ft. Attached Antenna

MODEL 50X2—With Handle
5 Tubes: A.C./D.C. Superhet: Tuning Range 540-1720 kc.: Powerful Speaker: "Aero-Vane" Loop: Ivory Finish Plastic Cabinet—10″ x 6″ x 6½″. U. L. Appd.
Model 50XH2 As Above—No Loop, but with 20 ft. Attached Antenna

MODEL 50XC—With Handle
5 Tubes: A.C./D.C.: 540-1720 kc.: "Aero-Vane" Loop: 10″ Bright Plastic.
50XC1—Red with White Trim
50XC2—Turquoise Blue, White Trim
50XC3—White Onyx
50XC4—Brown Onyx, Amber Trim

MODEL 50XW
5 Tubes: A.C./D.C. Superhet: Tuning Range 540-1720 kc.: Powerful Speaker: "Aero-Vane" Loop Antenna: Beautiful Selected Walnut Veneer Cabinet with Exquisite Carved Overlay—Size 11⅛″ x 6½″ x 5¾″: U. L. Approved.

MODEL 52XAH1
5 Tubes: A.C./D.C. Superhet: 4 Automatic Push Buttons: 2 Bands: Tuning Range 540-1720 kc. and 5650-18,000 kc.: 5″ Speaker—20 ft. Attached Antenna: Brown and Ivory Plastic Cabinet—11¾″ x 6⅞″ x 7¾″: U. L. Approved.

MODEL 56XA1
Tubes: A.C./D.C. Superhet: 4 Automatic Push Buttons: Tuning Range 540-1720 kc., includes 1st Police Band: Extreme Sensitivity: 5″ Speaker: "Aero-Vane" Loop: Brown Plastic Cabinet—11¾″ x 6⅞″ x 7¾″: U. L. Approved.

MODEL 56X1
5 Tubes: A.C./D.C. Superhet: Tuning Range 540-1720 kc., includes 1st Police Band: Extreme Sensitivity: Razor-Sharp Selectivity: 5″ Speaker: "Aero-Vane" Loop: Brown Plastic Cabinet—11¾″ x 6⅞″ x 7¾″: U. L. Approved.

MODEL 56XAW
5 Tubes: A.C./D.C. Superhet: 4 Automatic Push Buttons: Tuning Range 540-1720 kc., includes 1st Police Band: Extreme Sensitivity: 5″ Speaker: "Aero-Vane" Loop: Walnut Veneer Cabinet—Size 12″ x 7″ x 6¾″. U. L. Approved.

MODEL 60X1—With Handle
6 Tubes: Broadband R.F. Amplifier Stage: A.C./D.C. Superhet: Tuning Range 540-1720 kc.: 5″ Speaker: "Aero-Vane" Loop Antenna: Wave Trap: Gold Trim Brown Plastic Cabinet—Size 11¾″ x 6⅞″ x 7¾″: U. L. Approved.

MODEL 60X2—With Handle
6 Tubes: Broadband R.F. Amplifier Stage: A.C./D.C. Superhet: Tuning Range 540-1720 kc.: 5″ Speaker: "Aero-Vane" Loop Antenna: Wave Trap: Ivory Finish Plastic Cabinet—Size 11¾″ x 6⅞″ x 7¾″: U. L. Approved.

MODEL 60XW
6 Tubes: Broadband R.F. Amplifier Stage: A.C./D.C. Superhet: Tuning Range 540-1720 kc., includes 1st Police Band: 5″ Speaker: "Aero-Vane" Loop: Wave Trap: Large Walnut Veneer Cabinet—13¼″ x 7⅜″ x 7″: U. L. Approved.

MODEL 60XA1—With Handle
Tubes: Broadband R.F. Amplifier Stage: A.C./D.C. Superhet: 4 Automatic Push Buttons: Range 540-1720 kc.: 5″ Speaker—Tone Control: "Aero-Vane" Loop: Wave Trap: Gold Trim Brown Plastic Cabinet—11¾″x6⅞″x7¾″: U. L. Appd.

MODEL 60XA2—With Handle
6 Tubes: Broadband R.F. Amplifier Stage: A.C./D.C.: 4 Automatic Push Buttons: Range 540-1720 kc.: 5″ Speaker—Tone Control: "Aero-Vane" Loop: Wave Trap: Gold Trim Ivory Finish Plastic Cabinet—11¾″x6⅞″x7¾″: U. L. Appd.

MODEL 40-60W—With Handle
6 Tubes: Broadband R.F. Amplifier Stage: A.C./D.C. Superhet: 2 Bands: Tuning Range 540-1600 kc. and 1600-3700 kc.: 5″ Speaker—Tone Control: "Aero-Vane" Loop: Cabinet of Walnut Veneers—11¾″ x 7⅜″ x 5⅜″: U. L. Approved.

MODEL 50W
5 Tubes: A.C. Superhet: Tuning Range 540-1720 kc., includes 1st Police Band: 5½″ P.M. Dynamic Speaker: "Aero-Vane" Loop: Cabinet of Selected Walnut Veneers with Maple and Contrasting Walnut Inlays—15″x9″x7″: U. L. Appd.

MODEL 52T1
5 Tubes: A.C. Superhet: 6 Automatic Push Buttons: 2 Bands—American and Foreign: Range 540-1720 kc. and 5650-18,000 kc.: 5¼″ P.M. Dynamic Speaker: "Aero-Vane" Loop: Walnut Veneer Cabinet—16½″ x 9¼″ x 8″: U. L. Approved.

MODEL 62T1
6 Tubes: 3-Gang Tuning Condenser: Tuned R.F. Amplifier Stage: A.C. Superhet: 2 Bands: Range 540-1720 kc. and 5650-18,000 kc.: 6″ Speaker—Tone Control: "Aero-Vane" Loop: Selected Walnut Veneer Cabinet—16½″ x 9¼″ x 9⅝″: U. L. Approved.

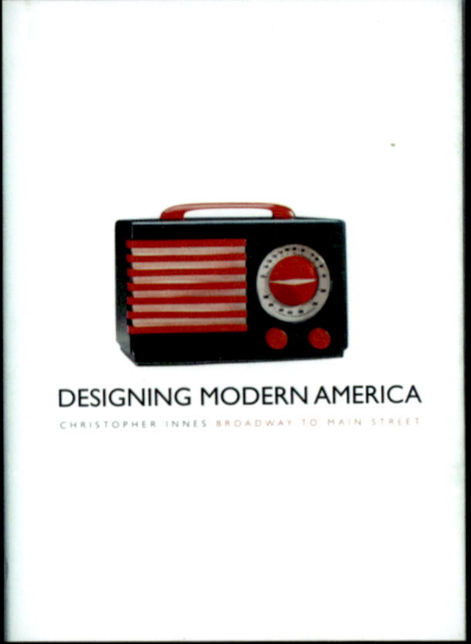

not cherished by radio enthusiasts whose predilection for technology saw them honor the chassis and valves rather than the cabinet. In fact, many in the radio world today struggle with the increasing value and collectability of colored Catalin and Bakelite radios in contrast with the plateaued desirability and value of the more common wood radios. Indeed, they often cost more to restore than they can be sold for.

Rather than this being the triumph of trash over treasure, it is a true reflection of the historical importance of these items and the fact that anyone who sees colored radios, whether collector or visitor, has a positive visceral response, usually one of surprise and awe at their individual and collective beauty. There is also an urge to touch and stroke the cabinets, which mimic all sorts of precious materials and their tactile nature seems to confound the fact that they are plastic, a material that today stands for things cheap, unfashionable, and ecologically inappropriate.

The chroniclers of Art Deco history disregarded these iconic representatives in describing the dynamics and history of this stylistic trend. It is clearly difficult to now retrofit the Deco radio into the evolution of the Art Deco style. One reason for the lack of importance attributed to the radio is that these were items mass-produced for the general public in the new style, not unique artifacts that are generally the cherished examples depicted in the Art Deco texts. But, I believe radio is significant and integral to the story. It is only recently that the radio rates a mention in serious books on Art Deco—one of the first being the catalog accompanying the 2005 Victoria and Albert Museum's Art Deco Exhibition in London.

Even the industrial designers who made their mark in the 1930s with cars, buildings, and trains rarely highlight the radios they conceived. A good example of this is seen in the book *Designing Modern America—Broadway to Main Street* (2005). In 300 pages on Urban and Bel Geddes, there is one paragraph mentioning the Emerson "Patriot" radio designed by the latter in 1940, yet a colored picture of this radio is the single feature of the cover of the book. Radio rates little importance in the text as a reflection of the efforts of Bel Geddes, but the publishers clearly identified the radio as the best image to convey the essence of the thesis and market the book. Perhaps the book reflects Bel Geddes hierarchy of his designs, or perhaps the author imposed a fairly standard view of what constitutes important elements in Bel Geddes' career. But, whoever chose the cover reflected the preferences of the twenty-first century and an appreciation of how much this particular radio design informs our tastes today.

In contrast, a very recent book, *Norman Bel Geddes Designs America* (2012), devotes a chapter to Domestic Media Designs and this includes a full description of Geddes work designing radios for Philco, Majestic, and Emerson. Belatedly, these commissions are gaining traction as epochal representations of Streamlined styling and will gain credence as important manifestations of the industrial designers' impact on the daily life of the community.

CHAPTER 4

RADIO—THE EARLY DAYS

In 1892, Nikola Tesla experimentally demonstrated the transmission and radiation of radio frequency energy and proposed that it might be used for the telecommunication of information. Guglielmo Marconi commenced transmissions of radio signals using a wireless system in 1895 and was awarded the initial patent for radio in Britain in 1897. In the same year, he founded The Wireless Telegraph & Signal Company and in 1898 opened the world's first radio factory near London. The company became a pioneer of wireless long distance communication and mass-media broadcasting. Much of the early use of radio was for ship-to-shore communication, but by the turn of the century there was worldwide interest in radio and many people experimenting with approaches to master the great potential of sound at distance.

The early 1920s saw rapid growth in the number of radio stations in most developed countries. The first radios introduced by pioneers such as Atwater Kent were called "breadboards," each being a simple, flat, wooden board with the valves and components mounted on top. Neither design nor cabinetry was important in this initial phase. Many of the early radios were crystal sets, which had no battery or amplification and required headphones. They generally could only receive signals from local radio stations and had difficulty in separating stations. Once radio sets with speakers became available, the crystal sets became more of an introductory novelty, appealing to the less wealthy, as they were cheap to build, and, as well, to the enthusiastic amateur rather than the average householder.

The vacuum tube was developed in the early 1900s and the earliest valve receivers appeared in the early 1920s with the distinct advantage of being able to power a loudspeaker. Tuned radio frequency (TRF) receivers had an array of knobs and switches, making tuning-in a radio station a major exercise. In 1925, the super heterodyne radio receiver was released—this being much more selective and sensitive with fewer knobs and dials, much like the radio of today. Mains powered receivers and battery eliminators appeared in 1926. Two General Electric researchers in the USA, Rice and Kellogg, patented the electrodynamic moving coil speaker in 1924

Astrophone crystal set c. 1923

Crystal set 1924

Valves 1926

Loewe OE433 1926

Valves and loudspeaker 1927

Philips "Permagnetic" Loudspeaker (1929)

and this quickly replaced the inefficient horn speaker that had originally developed with the gramophone.

Radio's popularity increased dramatically in the 1920s. The number of radio licenses in the UK grew from 15,000 in 1922, when the BBC began broadcasting, to 3,000,000 in 1930, 5,700,000 in 1934, and 9,000,000 at the end of 1948. It is worth noting that in 1948 some twenty-five percent of the UK still did not have mains electricity.

In the USA in 1925, nineteen percent of households had a radio, growing to forty percent in 1929, sixty percent in 1934. And, by 1941, ninety percent of families had a radio, listening to an average of three to four hours a day, and being the main source of news for millions of people. The number of radio stations grew from 5 in 1921 to 765 in 1940 and in that year, as well, radios were installed in over twenty-five percent of American cars.

Our story truly comes alive in 1930, but we need to start just a little earlier, in 1926 in Germany, with the Loewe company and their OE433 (see left), its little Bakelite molded case extending the routine use of Bakelite for switches and electrical components. Alongside its wooden counterpart, the OE333 (which was truly a wooden box with little in terms of design or finish), this elegant little base gives a sense of integration as well as functional appropriateness and in retrospect we see a hint of the future. Loewe also introduced in this model the first integrated valve and so it has technical as well as design importance. This model was also available with little delay in England, indicating the growing international passion for radio. It was powered by an external battery and then connected to either headphones or a loudspeaker.

The predominant fashion in the radio world of 1927 reflected Victorian or Edwardian influences, so when the Philips PCJJ electromagnetic loudspeaker designed by Louis C. Kalff was released that year in the Netherlands, it heralded a true paradigm shift (see right). It has been nicknamed the "shaving plate" or "flying saucer" and is a genuine example of creative industrial design. Philips was the first factory in the Netherlands to process Bakelite and in 1923 started with hand presses to produce Bakelite components for radios and electrical engineering, calling their version of Bakelite "Philite." Philips considered their new loudspeaker an object of art and at that time required the largest mold in the world to produce it in Bakelite.

In retrospect, we can see a novel and distinct break from both traditional materials and design. A circular, concave, Bakelite parabolic bowl stands on a metal foot and acts as a reflector with a smaller convex disk seemingly suspended in front of it. There is a sense of sound appearing from behind a solid façade and no evidence of the common speaker cloth, which was hidden behind the front disk that contained the motor and paper cone. Most speakers were standard brown, but due to the random addition of pigments there is enormous variation in the mottling, showing an addition of red and green and some with additional brass flecking. They were available in three sizes (2007 large, 2003 medium, 2015 small) and were marketed throughout Europe, England, and Australia, being extraordinarily popular till 1932. Kalff also designed the first Philips radio (2501), which was advertised together with the PCJJ loudspeaker. This coffin-shaped metal box was ostensibly in the modern "new realism" style, but it looks old-fashioned and lacks the aesthetic impact of the PCJJ loudspeaker. In 1930, the Philite factory employed some 750 people, although this dropped to 450 in 1933 due to the Depression.

In 1929, Philips released the Kalff designed 2113 "Permagnetic" (see left), a moving coil speaker that introduced rudimentary baffling by enclosing it in an open-backed cabinet. The floor model is a diamond-shaped metal surround with a Bakelite face, all seated on metal legs. The actual speaker is a small, projecting, seven-sided louvred metal grill in the bottom half of the Bakelite front. Nicknamed "The Martian," this attractive piece (attributed to Louis Kalff) is resonant of the futuristic themes in the film *Metropolis*, released a year or so earlier. The PCJJ cost £5–£7, equivalent to the weekly average wage at the time. The "Permagnetic"—at £19 for a loudspeaker—was obviously marketed to a section of the community with much higher incomes.

But, while Philips was doing very well commercially with their loudspeakers, this was a watershed period for radio. The USA, UK, and Germany all have been touted as having introduced the first integrated tabletop radio with receiver and speaker all contained in a wooden cabinet and utilizing a wall socket for electricity. It is hard to know if the advent of the tabletop radio was a commercial response to the Depression or the next incremental stage in radio development. There is little doubt that this new wave of radios (designated variously as midget, compact, tabletop, mantel, or portable) was significantly cheaper than wooden consoles and infinitely more affordable to consumers who were struggling to put food on the table. Today, we would hardly consider these monumental cathedral style radios as small, knowing as we do about transistor radios and iPods. But, at the time, they were a quantum reduction in the size of radio, with little or no loss in performance, while still offering huge potential for expansion into new homes, into new rooms, and for personalized as well as family listening.

In the USA, Philco and Ecophone were the frontrunners; in Germany, AEG and Nora; and in the UK, Ekco, Murphy, and Kolster Brandes. In 1930, all these companies had tabletop radios in production, but the German Nora model K3W released in 1929 appears to be the first true tabletop radio. As we will see, this is the company that heralds the next major step in radio manufacture. The Nora K3W is prototypically simple in design, with a tombstone outline, yet without any of the typical gothic flourishes and embellishments that were routine on contemporary radios. The front fascia below the plain speaker fabric is made from Bakelite (or Arbolite) and contains three knobs and a small rotary dial. The base and feet display a more traditional curved border and this motif is retained in later Nora models in both wood and Bakelite.

One year later in 1930, with little fanfare, the Nora company in Germany released the all-Bakelite "Sonnenblume" (sunflower) in two versions. It came as a traditional speaker to go with its W3 radio receiver and, most significantly, as a fully fledged independent integrated tabletop radio powered by standard household electricity. Here we have the genesis of the modern mantel radio with its cabinet made from plastic and in a style that speaks of utility and modernity. The cabinet still retains a tombstone orientation similar to the K3W, but in the transition to Bakelite it discards its patrimony and stands prototypically smooth and lustrous. There is some sense of naked pride in the glossy, polished, continuously curving surface devoid of joins or hard edges. It has a sculpted thin profile compared to most radios of the time and this adds to the sense of minimalism and refined styling. The use of the flower and the name symbolizes the blossoming of a new form. The petals of the sunflower dominate the cabinet face, but in their stylized arrangement allow the speaker grill cloth to form much of their surface. The rotating dials are simple and unobtrusive, leaving an uncluttered and undecorated surface. The knob forming the pistil of the flower is cleverly camouflaged. The Sonnenblume is a complete radio, its cabinet molded from Bakelite and ready for mass production. It is unquestionably a beautiful example of European Art Deco in a functional and commercial entity. Of course, by 1930 the rise of fascism in Germany was well underway and changed the future of radio design and usage in Germany. We will return to follow the fortunes of the Nora radios in Chapter 8.

England was a pioneer in the evolution of radio, but modernity seemed to struggle against the obduracy of tradition, particularly in the tastes of the British consumer. There are three companies that monopolized the radio scene in the 1930s and 1940s, Ekco, Murphy, and Kolster Brandes. These companies seemed, as a group, to embrace modern design in the early 1930s with Murphy restricting itself to wood while Ekco and KB immediately opted for the new synthetic plastics using both phenol- and urea-formaldehyde. England was not far behind Germany in the use of Bakelite for radio cases. In fact, the Ekco Company had cases for their models 312 and 313 made by AEG in Germany in 1930 before the imposition of high import duties on radio cabinets, in 1931, prompted Ekco to establish its own Bakelite molding shop adjacent to

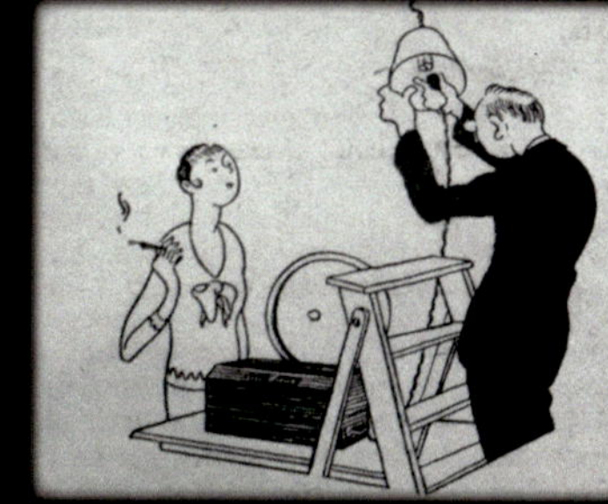

Add mains electricity 1930

Philips 820A radio and PCJJ loudspeaker

Nora "Sonnenblume" 1930

Art Deco radio 1932

their new plant in Southend-on-Sea. By 1932, Ekco had over 2000 employees, most of them women who worked on the factory floor.

Under the inspired leadership of its owner, Eric Kirkham Cole, Ekco not only embraced the use of Bakelite for their radios, but co-opted the best designers in England to maximize the potential of the new material. The first in a wonderful series of mantel radios was the RS3 in 1932 (p. 118), a large Bakelite molding designed by J.K. White. The cabinet was Deco-influenced—upright with stepped shoulders—while the distinctive willow tree pattern in the loudspeaker grill clearly harks back to Art Nouveau style. In the same year, Ekco's RS2 used a similarly shaped, but smaller, Bakelite cabinet, but the front has been simplified with the exposed speaker grill interrupted by two vertical Bakelite bars extending up from either side of the small window dial. This radio case was probably the first in the UK to entirely embody Art Deco styling in the plastic medium. Cole gathered around him some of the best architect/designers in England (Coates, Black, Chermayeff, Collins, White) and instituted a competition for the design of a new radio. Wells Coate's AD65 was the winner, but many of the designers saw their creations become commercial realities.

The Wells Coates' designed Ekco AD65 of 1934 (pp. 139-141), perhaps more than any other radio of its time, reflected machine-age engineering rather than traditional hand-crafted furniture. The design is unashamedly modern, almost completely devoid of decorative elements. The chrome vertical grill elements over the speaker cloth and the half round chrome bar on the dial add emphasis rather than ornamentation. The black version also came with chrome inserts on the front of the knobs and imparts a subtlety and gravitas that no other radio has ever attained. This reflects the certainty of the design, the homogenous grafting of shape and function, creating an entity that supersedes its domestic setting and the everyday nature of the media it radiates. Sometimes it takes many decades before we can look back at a design and recognize its originality and significance. Clearly the British public saw something special in this round, plastic radio that transcended their ties to the past to the extent that the core round design maintained popularity through five different models and for some ten years. The green version of the AD65 is extraordinary for its rarity, colored radios being completely ignored by the British public in the 1930s. It is not known how many were made, but today only four decent examples are known to exist. It stands resolute in defiance of its dismissal, its jade-like surface mocking its plastic legacy. It is untouchable, yet its smooth, rounded surface demands to be stroked, gently and with reverence.

The Jesse Collins designed Ekco AC97 of 1936 (p. 169) is one of the largest Bakelite radios ever produced (21"H x 13"W x 8"D) (54cmH x 33cmW x 20cmD). In brown it is a lump of a radio, but in black with white knobs and escutcheon it assumes a persona. Nicknamed "Robby the Robot," it is one of the few large radios with human-like features and is always spoken of by owners in affectionate terms. Actually, the largest Bakelite molding ever made is a coffin that visually mimicked dark wood and was probably cheaper to produce. Nevertheless, it was a commercial failure and the single example has been laid to rest in the Science Museum in London.

One of the first English Bakelite radios is the Kolster Brandes (KB) "Masterpiece" (1931; pp. 23, 107), which was basically a speaker hinged to the receiver forming a square box when closed. The base contains two valves and the knobs, while the lid is the speaker. The exposed wires extending from the base to the lid signal its links to the crystal sets of the 1920s. KB supplied the entire production run of 40,000 to the Godfrey Phillips tobacco company, who gave them away to customers in exchange for 500 coupons collected when purchasing a similar number of packets of Best Deck Virginia cigarettes.

The Murphy company produced a number of wooden tabletop radios from 1931 onwards using the cabinet designer Richard Drew Russell. His radios all maintained a simple primary box shape and limited decoration reflecting a modern approach to design. The A3 of 1931 was one of the first mass produced mains radio in the UK and also the first using walnut veneered plywood. The A8 of 1932 was Murphy's first superhet receiver, the first cabinet in the UK to have a landscape profile, and the first radio in the UK to have automatic volume control.

Major radio manufacturers in the USA were, at first, slow to incorporate modern styling and materials in their radios. The initial tabletop radios were cathedral shapes and with traditional gothic moldings. Philco, in 1930, on the basis of sales of its wooden cathedral-shaped Model 20 Baby Grand mantel radio went from the third largest radio manufacturer in the USA to number one by the end of that year.

In 1931, the International Radio Corporation released the tiny Kadette "H" series in a Bakelite case (p. 109). It was popular with consumers and contributed to the company being one of the few radio manufacturers who were profitable during the Depression. Presented in a rectangular case, the front was very gothic in design, but this radio was an important forerunner of American radios, which made them different from their European and Australian counterparts. This first radio without a transformer allowed a huge reduction in the size of the chassis, which meant the cabinet was only 7.5" (19cm) wide by 5.5" (14cm) high and 4" (10cm) deep (one third of the height of the Philco 20 at 17.5" (45cm) tall). This was possible because mains electricity in the USA runs at 110 volts, whereas Europe and Australia have 220–240 volts. The larger current would burn out the workings in a radio without a transformer.

The majority of European mantel radios of the 1930s and 1940s were consistently larger than 12" (30cm) high. Australia produced a number of smaller radios at about 6" (15cm) height, but the weight of these (due to the extra transformer) was almost three times that of American radios of the same size. Not only did the Kadette "H" indicate the future size of Catalin radios in the second half of the 1930s but introduced color—other than dark brown or black—for the first time in the world.

America's Air King radios of 1933 (pp116-119) were a radical step in a new direction for radio, presenting a fully formed vehicle for modern styling, color, and new materials. The Air King utilized the new synthetic plastics to mass produce a cabinet, which, in channeling the skyscraper, brought the essence of modernity to the radio world and to the consumer. It was offered with two distinct shapes (single knob or three knob), three different inserts (Egyptian scene, map, and clock) and 11–13 different colors and finishes. This smorgasbord of choices for the buyer was in stark contrast to the limited variety in wooden models. Additionally, the Air King was also light and portable, presaging the extension of radio into every room in the house and the expanded versatility of the tabletop model.

Australia's entry to tabletop radios came with the spectacular Bakelite AWA Radiolette C87 (1932) and C104 (1932/33; p. 111). Both of these radios share the same cathedral cabinet with the C87 having two knobs and dramatic red/brown mottling, while the C104 (introduced within 12 months of the C87) had three knobs, somber brown coloring, and was adorned on the front with contrasting mottled decorative panels on either side of the center. There is a variant of the C87 with a GE logo within the molding (rather than the AWA logo) and this indicates perhaps an American association with AWA at this time. The shape of the cabinet is traditional cathedral and the front is Gothic with overtones of Nouveau and a hint of Deco. However, in the new material, Bakelite, the radio was truly transitional, the hard edges smoothed off and presenting a much more cohesive and integrated object than its wood predecessors.

The AWA Fisk Radiolette of 1934 (pp. 144, 145) is the Australian equivalent of the US Air King, being a novel shape and offered in a variety of colors. Nicknamed the "Empire State," this radio is a solid and dignified skyscraper in which the architectural elements are cleverly replicated and integrated into the plastic cabinet. The speaker is well hidden behind the vertical flutes and the dial is visible, yet unobtrusive. While clearly modeled on the Air King, it is not a carbon copy and is respected by collectors worldwide. In 1935, AWA released another Fisk Radiolette, which has been called the "fret'n foot" (pp. 162-164). It is a bold and bulky mantel radio and perhaps with too many different features in the grill (Chinese lattice) and feet (lions paws), but is a world-class example of modern styling and one of the first radios with cabinet elements in contrasting colors.

Ekco AD65

KB "Masterpiece"

Shop display with Ekco AC97

Kadette "Junior"

Nora "Sonnenblume"
Germany 1930

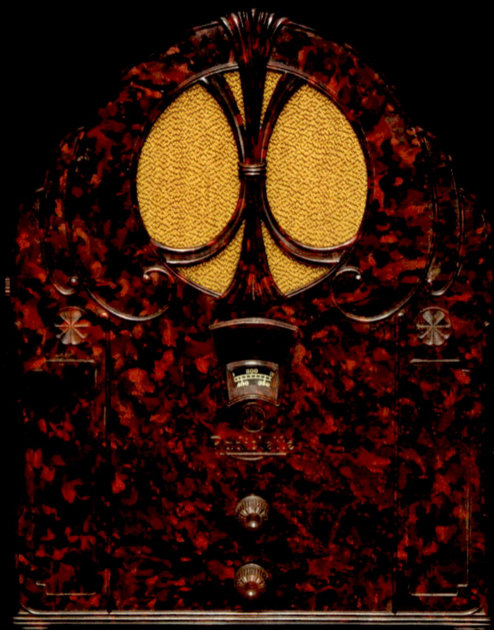

AWA C87
Australia 1932

Ekco RS3
UK 1932

Air King 52
USA 1933

CHAPTER 5
THE RESIN REVOLUTION

It has been suggested that some seventy-five percent of all things made today are from resins ... better known as plastics. And while we talk reverently about the Bronze Age, the Iron Age and the Atomic Age, for some reason we make no reference to the incredible impact that plastics made to the life of people on this planet in the twentieth century and continuing on into the twenty-first century. In so many instances, replacement of traditional metals and woods with plastic has meant a significant reduction in the cost, complexity, and weight of objects. While we generally look at the effect of plastics in the Western world, a simple example of a domestic application in so-called developing societies illustrates the transformation. The advent of the plastic bucket has dramatically altered the lives of women and children in many parts of Africa. This has transformed the way of transporting of water in a country where this daily task has often been very arduous.

At the end of the nineteenth century, there were many applications for non-metal materials, but all had an organic basis. Natural and semi-synthetic plastics were used extensively for jewelry and small domestic items. These included celluloid from plants, vulcanite and gutta-percha from trees, tortoiseshell and casein from animals, and shellac from insects. They all had significant weaknesses with variable strength and high flammability and there was a pressing need for a hard synthetic material. For example, due to the reduction in ivory obtainable from Africa, there was pressure for a non-ivory billiard ball.

Bakelite was the first synthetic plastic and the subsequent spread in the early decades of the twentieth century of commercial uses for synthetic plastics was an important part of the gathering momentum of the machine age and mass production in Western society. From an economic point of view, it was cheap to produce in bulk, and reduced costs by the relative ease of finishing. It was also inorganic, with dependable supply and not reliant on flora or fauna. From a technical standpoint it was lightweight, dimensionally stable, and precise. From a consumer's vantage, it was a symbol of modernity and the future of technology at a time when the Depression and impending war had undermined the confidence of the community. Finally, it was a material that allowed the designer a new medium to fully express the modern styles and forms that symbolized the new age of speed and utility.

Plastics have an extraordinarily diverse range of properties and technologies that allow for multiple and economical applications. This versatility has somehow degraded the appreciation for the benefits of this class of materials to the extent that the

word "plastic" (originally from the Greek "*plastikos*" meaning pliable) has become a synonym for something cheap, artificial, inconsequential, and unreliable.

The world of early plastics was complicated by the similarity of generic, company, and product names, and it is always mind-numbing to try to clarify subtle differences. The plastic used for radio cases in the 1930s and 1940s are in the main thermoset resins based on two chemical compounds: phenol-formaldehyde and urea-formaldehyde. Some additional radio parts were made from cellulose acetate.

The original phenolic resin, Bakelite, was invented by Leo Baekeland in 1907. This is a thermosetting plastic derived from two by-products of coal and oil: phenol and formaldehyde. Mixing them together under moderate heat produced a viscous resin that was then allowed to cool. The hardened mix was crushed and fillers of cotton flock, wood flour, and paper were added, and the mixture then molded into permanent shapes using heat and pressure. Colors were limited to dark hues.

Substituting urea and cellulose fiber for phenol produced urea-formaldehyde and this was developed in the UK in 1925. Moldings with urea-formaldehyde were able to be produced in much lighter colors than Bakelite. Brand names were Beetle, Banadalasta, and Beetleware; in the USA the Toledo Scale Company developed and introduced Plaskon in 1931. The terms beetle in the UK and plaskon in the USA are used as generic descriptors of urea-formaldehyde molded resins.

Catalin is a phenol-formaldehyde resin, which is produced by casting rather than molding and is described as a cast resin. Catalin was introduced in 1927 by the American Catalin Corporation with brand names such as Catalin, Fiberlon, and Opalon. It was transparent and filler-free. This allowed the introduction of infinite colors and color variations. American Catalin built radio cabinets for FADA, RCA, and GE, while the Bakelite Corporation produced the RCA Little Nipper in 1938. Fiberloid made the Emerson AU190 from Opalon cast resin. Ortho Plastic Novelties produced the Emerson "Patriot" using Opalon. All phenol and urea-formaldehyde resins are thermosetting and once formed under pressure cannot be reused.

Cellulose acetate is a thermoplastic resin (which can be softened and will reharden) and was the first plastic capable of being injection molded. A major brand was Tenite, used for grills, knobs, dial covers, and pointers.

Leo Baekeland is a fascinating character and the father of modern plastics. Baekeland was a Belgian chemist who went to the USA in 1890 to better his circumstances. He invented a photographic process called "Velox," which was very popular with amateur photographers. He arranged a meeting with Kodak's President, George Eastman, in 1899, where he was determined to take no less than $25,000 for the product. He walked away with $1,000,000 and the future was his to explore.

He noted that the USA was importing over one million pounds of shellac each year and decided that what the world needed was a synthetic shellac. In 1907, his experimentation culminated in Bakelite, the first synthetic plastic. He named his invention "Bakelite" after himself and in 1924 reached the level of accomplishment where he was seen on the cover of *Time Magazine*. His boast that Bakelite was "the material of a thousand uses" was a wonderful marketing strap-line, but ultimately a hyperbolic understatement given the plethora of uses to which Bakelite and its offshoots were put.

The major features of Bakelite were that it was hard and flame resistant. At that time, it was considered a benefit that it could not be melted back to its original and reused (i.e., thermoset as against thermoplastic)—a property that today would be unacceptable given the amount of redundant plastic in our lives and the ecological benefit of recycling.

At first, the objects formed of Bakelite were relatively small; its major application was for electrical insulation, plugs, light switches, batons, etc. However, Baekeland was prescient and the range of uses ballooned to include kitchenware and barware, and, of course, billiard balls.

MOLDED PHENOLIC RESINS
e.g. bakelite

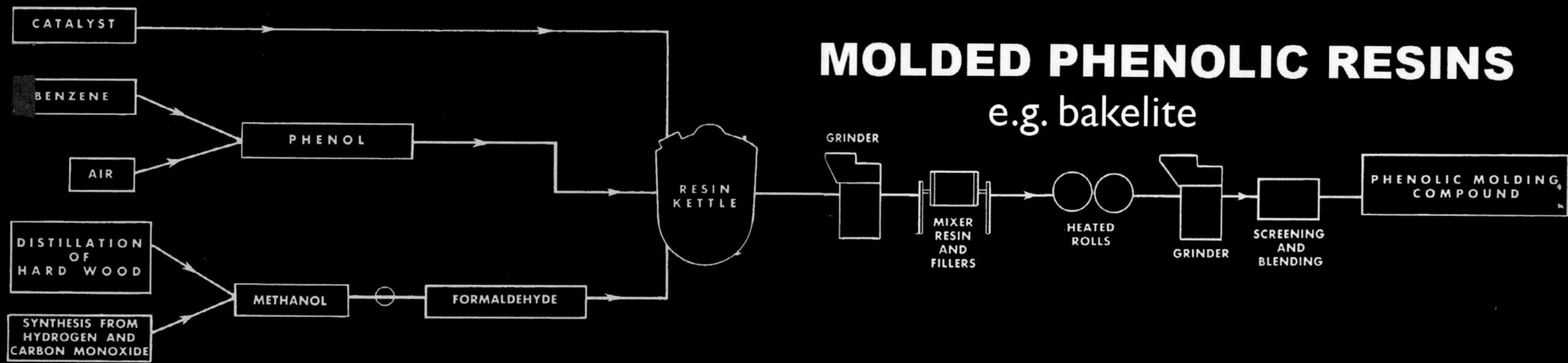

CAST PHENOLIC RESINS
e.g. catalin

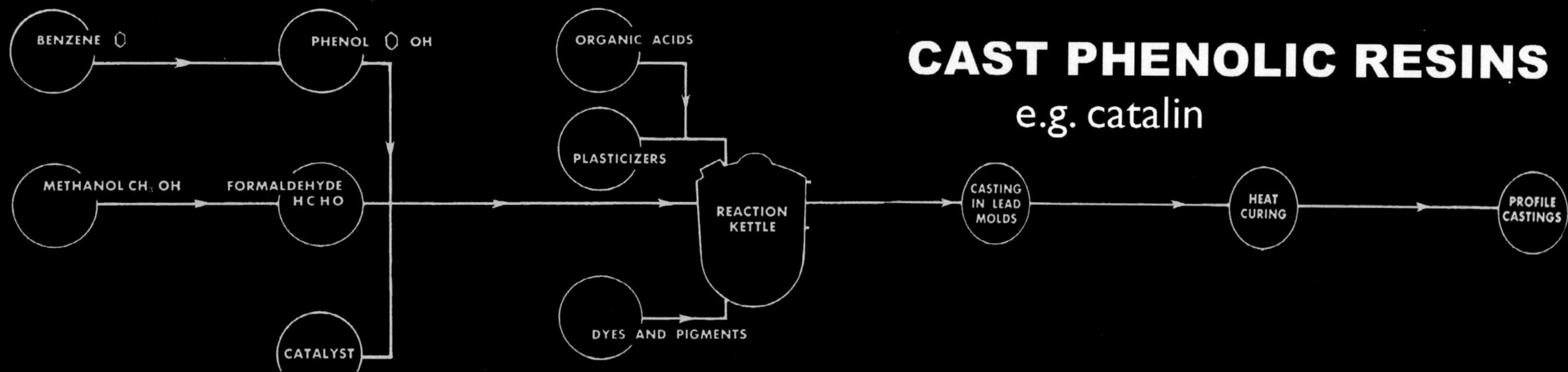

MOLDED UREA RESINS
e.g. beetle, plaskon

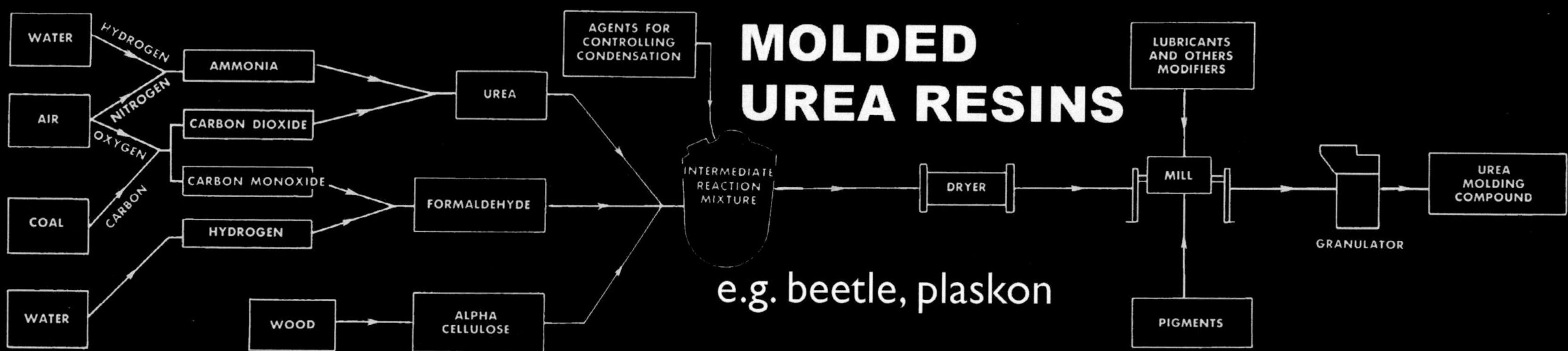

Bakelite was dark colored, with any variations dependent on the filler and pigments being added. In some ways, this was valuable for its integration into the life of the early twentieth century as so much of the structures of houses and buildings were made of wood. Bakelite was initially a small contributor that fitted well with the dark tones of Victorian and Edwardian life.

Within this limitation, producers attempted to mimic the colors and variegation in different woods. Thus, some Bakelite might be more a walnut or a mahogany and occasionally even black. Spreading pigments throughout the mix prior to molding created mottling that could mimic some of the more exotic and expensive woods. During World War I, Bakelite was used in battleship insulation and for artillery shell casings. Baekeland promoted a Bakelite propeller for use by combat aircraft and after the war this was adapted to become the impeller in automatic washing machines. While most uses of these resins were for utilitarian objects, there are many examples of beautiful objects that reflect the versatility of these materials and the potential when talented designers become involved. One very rare piece is the signed Bakelite powder box c. 1930 in a dark red/orange color with the lid displaying a delicate setting of embossed cherries (p. 29). The Carvacraft bookends (UK c 1945) provide another beautiful example, designed by the English silversmith Charles Boyton before World War II, which have an organic teardrop shape combined with classical Art Deco linear elements (p. 29).

The use of Bakelite for radio speakers commenced in 1927 in Holland (Philips) and for cases in 1930 in Germany (Nora) as production methods for large moldings improved. Bakelite radio cabinets were first produced in the UK in 1931 (Ekco and Kolster Brandes), the USA in 1931 (Kadette), and Australia in 1932 (AWA). The color was generally brown, but variations extended to black, dark reds and greens, and any of these could contain mottling.

In the UK, the British Cyanide Co. introduced urea-formaldehyde resins in 1925 for tableware in bright and hygienic looking colors under the product names of Beetle, Beetleware, and Bandalasta. By 1927, the American Cyanamid Company had acquired the rights to produce Beetleware in the USA. Molded radio cabinets using urea-formaldehyde were first seen there in 1931 (Kadette), and in 1934 in the UK (Ekco) and Australia (AWA). British Industrial Plastics, the successor to British Cyanide, supplied urea-formaldehyde molding materials to both Ekco in the UK and AWA in Australia.

Philips in the UK produced a series of radios in 1932 (e.g., 930C, 830A) using a material called Arbolite, the precursor of what we know today as "formica." Arbolite is formed by fusing layer upon layer of paper, cellulose, and phenol resin. This was pressed under huge pressure to form sheets and the outer layer then coated in a smooth finish. A similar, earlier form of the material can be seen on the front of the Nora K3W in 1929 (p. 29).

The American Catalin Corporation was founded in New York in 1927 when the patent on Bakelite expired. Catalin was the basis for most of the highly colored plastic jewelry in the late 1920s and 1930s with the company offering more than 20 different colors. Catalin was first used for radio cases in 1936 (FADA) and was never used for radio cases outside the USA. In 1942, the Catalin Corporation suspended production of commercial items to concentrate on wartime production. Newer petroleum-based plastics gained preeminence after the war. There were three major manufacturers of cast phenolic plastics in the USA: Catalin Corp., Bakelite Co., and Fiberloid (later Monsanto). A smaller company called Marblette made two castings for Emerson radios. Although the Bakelite Company was formed by Baekeland in 1911, it did not restrict itself to molded resins and after 1927 offered a cast phenolic with even more color variations than those from the Catalin Corporation.

By the mid-1930s, a burgeoning group of companies in the USA and UK were manufacturing synthetic resins, forcing down the price of Bakelite and Catalin. There were significant differences in the commercial approach to molded as compared to cast cabinets. The very significant cost of set-up for machinery in the case of the former assumed a large volume of production and this allowed the units to be sold at a cheap price. Also, molded cabinets required minor post-production polishing and cleaning, which reduced the labor costs.

Catalin on the other hand was a difficult material to work with and although the cost of setting up the molds was far less than for Bakelite, there were substantial numbers damaged and subsequently discarded. The finishing was labor intensive, increasing the per unit cost. Thus Bakelite and urea-formaldehyde radios are generally cheap and found in large numbers, whereas Catalin radios had small production runs and were considerably more expensive.

During World War II, there was extensive development of both old and new resins, with new materials and new production methods being fast-tracked for the war effort. Most of the radio companies in the UK, USA, and Australia halted or reduced production of commercial radio during the war as they reoriented their efforts for war-time communications. Many of the businesses failed to regain their commercial impetus after the war, some of the larger companies diversifying into other areas of communication such as television. Few still exist today: the survivors include Philips, Motorola, RCA, HMV, Siemens, and Telefunken.

The generic use of the term Bakelite today generally embraces both molded phenol- and urea-formaldehyde thermoset resins, although strictly speaking it should be limited to the former. The terms resin, polymer, and plastic are relatively interchangeable and include thermoplastics, thermosets, and elastomers. Polymers such as epoxy and phenolic are thermosetting plastics, whereas nylon, polyethylene, and polystyrene are thermoplastic. Plastics such as acrylic, polystyrene, and PVC were developed in the 1930s, refined, and expanded during the War and boomed in the 1950s. The war was also a conduit for better production methods and, thus, in the early 1950s, the Bakelites and Catalins were superseded by more modern, cheaper, and durable recyclable plastics.

Carvacraft Bookends UK c. 1945
John Dickinson Co.
Designer: Charles Boyton
Catalin

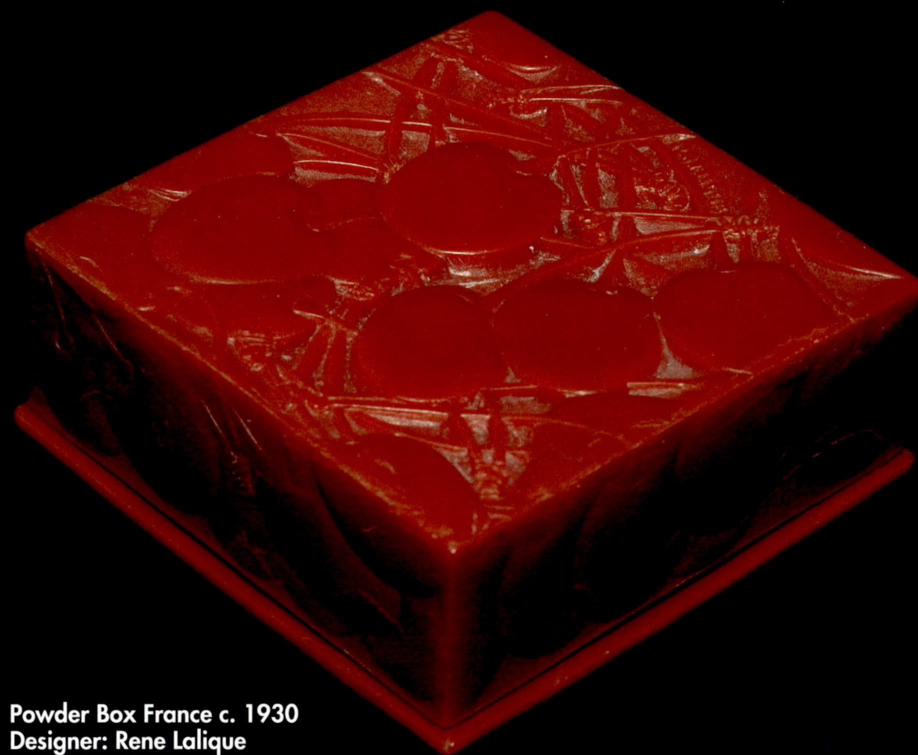

Powder Box France c. 1930
Designer: Rene Lalique

CHAPTER 6
NEW STYLES

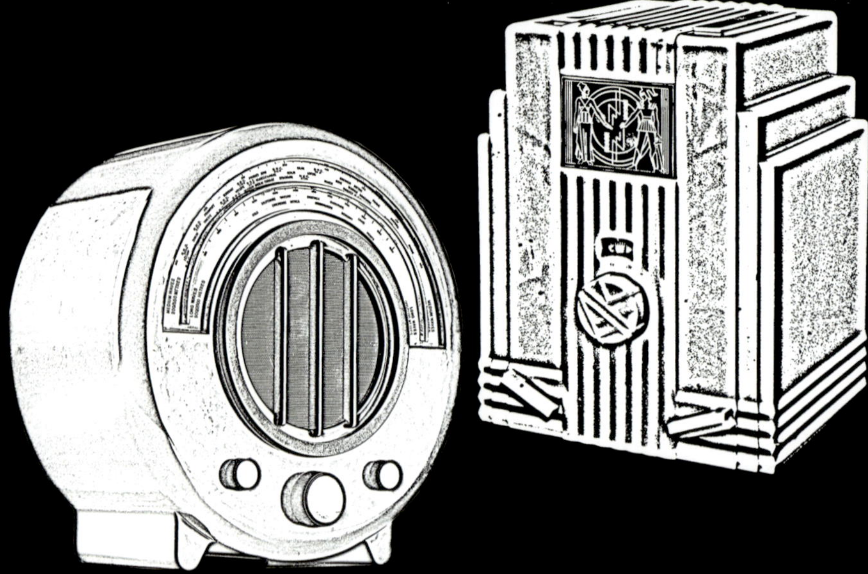

Radio stands as a unique twentieth century product—a novel technology with no outline inherited from earlier objects in previous centuries. In many ways, the mantel (or tabletop) radio was an amorphous, embryonic form only limited by the size of the valves it had to contain and otherwise ready for its essence to be clothed in new shapes and materials. The radio console owes something to the gramophone cabinet, but is far less significant than the car's debt to the horse and buggy.

Today the term Art Deco slips off the tongue as an easy descriptive phrase for an individual item or of a broad stylistic movement. Most of us know what we mean when we call something "Deco." Actually defining its features is more complicated because the term itself is a composite. The expression Art Deco was coined in the 1960s by the British art historian Bevis Hillier to describe the new styles of the 1920s and 1930s and is a contraction taken from the title of the World's Fair (*Exposition internationale des arts décoratifs et industriels modernes*) held in Paris in 1925 (see left). At this event, the most avant-garde of architecture and applied arts were showcased, but was essentially a European interpretation of decorative arts, as the Americans declined to present at the Exposition, claiming they had nothing new to offer. So the examples of the new modern style exhibited at the Paris Exposition reflect an aesthetic that evolved mainly in Europe from Art Nouveau and the Arts and Craft movement. The high point of this movement is the furniture of Emile-Jacques Ruhlmann, which embodies elements of classicism and resonances of Art Nouveau, albeit within the framework of elegant and functional styling. This design ethic is generally applied to individual items with an emphasis on craftsmanship and detailing and a sense of lavish and opulent decoration.

Missing in this view of Art Deco is the impetus of American Streamlining, which became a major design focus in the 1930s and informs much of our appreciation of Art Deco today. The designation "Moderne," "Streamlined Moderne," or "Machine-age" reflects the impact of American industrial design in architecture, transport, and virtually all forms of life-style in the 1930s. This approach is quite different to classical European Art Deco, with the practice of the artisan and his tools commissioned by the elite replaced by the designer working through the factory and machine to reach and benefit the whole community. In fact, for some, Streamlining was a response to Art Deco and, as well, to the difficult economic period during and after the Depression. Streamlining heralded an exciting new future with excellent design for all. Ornamentation was limited to simple lines and curves. Exotic materials were replaced by cheaper mass-produced components such as glass, cement, and plastic.

What underpins American Streamlining are the design movements of Europe such as the Bauhaus in Germany, De Stijl in the Netherlands, Italian Futurism, Russian Constructivism, Wiener Werkstatte in Austria, French Cubism and Dadaism in Switzerland. Elements of these styles added solidity and intellectual rigor to the ephemeral decorative nature of European Art Deco but all endorsed the machine and supported manufacturing technologies as repositories of art and style. Streamlining was more about style than ideology and the industrial design footprint extended broadly to the consumer via mass production. The emigration to America in the 1920s and 1930s of many of the leaders of these movements added to the drive for a break with tradition and a look to the future that attended the Streamline style. Additionally, with both Art Deco and Streamlining, many of the motifs were influenced by the artwork of indigenous groups and ancient civilizations. The difference in style is seen in the graphics of the posters from the 1925 Paris Exposition (p. 30) and the 1933 Chicago World's Fair (see right)—clearly there was a move from classical elements to the distinct lines and angles of American industry, architecture, and transport.

The term *"modernism"* adds more complexity to this assessment as it seems for some commentators to be an overriding movement in art, architecture, design, and literature, beginning in the late nineteenth century and continuing to the 1950s. While there is a sense of the rejection of tradition, classicism, and decoration in favor of abstraction and the power and potential of industrial and machine age technology, it is perhaps too wide-ranging, including all and signifying little. While embracing what is new can be avant-garde, the new is always shifting temporally. Style movements that describe themselves as "new" or "modern" can over the course of decades become historical anomalies, becoming everything but novel. The need for those who create to distance themselves from their antecedents and their peers is only matched by the penchant for those who follow to categorize and label. Perhaps all that can be said is that the evolution of design is a punctuated continuum; long periods of gradual modifications, interspersed with occasional major deviations.

Two loudspeakers of 1927 give a wonderful illustration of the variations in modern design occurring in Europe at that time. The beautiful French Hervor "Minilux" loudspeakers (pp. 92, 93) are quintessential examples of French decorative art reflecting the modernist touch while maintaining traditional values. They are tiny and with minimum performance; they are virtually handmade with celluloid, a fragile and somewhat unsuitable organic plastic mimicking the more expensive mother-of-pearl and tortoiseshell; the ornamentation ranges from the face of a flapper to idyllic pastoral scenes to abstract art. They are beautiful rather than functional. The Dutch Philips PCJJ Loudspeaker (pp. 90, 91) is unquestionably modern with futuristic lines and eschewing all ornamentation. They were mass-produced in Bakelite and came in three sizes, using a circular form with convex and concave segments. Their colors are dark, reflecting the limits of the material, but with subtle variegations. They are the essence of functional utility.

The Philips speaker sold in the hundreds of thousands all round the world, the Minilux is barely known outside of France, few were sold and just a handful have survived. The Minilux is a lovely art object, but the radio world needed a thing of beauty that also allowed for cost-effective mass-production and the Philips speaker showed the way.

In the 1930s, the view of design shifted from being the preserve of the elite to a more democratized access through machine and industry. There was also a rejection of ornate and unnecessary embellishment in architecture and decoration and an assimilation of technology and speed combining to provide a new future. Particularly in the USA, the goal of designers was to improve

people's lives by enhancing and reshaping consumer products. Many of the books describing Art Deco are clear on this point but then concentrate on highlighting the decorative aspects—the most exquisite furniture, metalware, sculpture, ceramics, artwork, and jewelry—almost all of which was handmade or produced in limited quantities. While the designer and artist are well represented, the industrial designers of the early 1930s are often ignored. Perhaps the academic aficionados find the meeting of art and industry a little low-brow, or reject the idea that elegant design can be reflected in mass-produced commercial items accessible to the whole community.

The instigators of Streamlining (Raymond Loewy, Walter Dorwin Teague, Norman Bel Geddes, and Gilbert Rohde) actually regarded Art Deco as posturing, falsely modern, and dishonest. The first three of these four eminent industrial designers are significantly represented in this book, but they would probably be confused at the convergence of Streamlining and Art Deco, which they saw, at the time, as not only different, but incompatible. Their designs for radio cabinets clearly are reflective of the modern style they were promoting. This gives further support to the thesis that the radios presented in *DecoRadio,* through their form, their core materials, and their social utility, signal an important and early role in the evolution of the Moderne and Streamlining movement in the USA and then throughout the world.

Notwithstanding the differing definitions and examples of the styles of the time, eighty years have softened the hard edges of disputed design ideology. Today, Art Deco is deemed a style with elements of Streamlined (or stripped) classicism; geometric, angular, and symmetric features; as well as elements of speed and the machine-age. The radios described here undoubtedly reflect Streamlining or Streamlined Moderne movements that today are considered an integral foundation of the style we all recognize as Art Deco.

To promote these radios to a higher status in the Art Deco hierarchy, the evidence base must be profound and compelling. Is it enough to point to the quality and quantity of the listed industrial designers and display the abundance of beautiful radios to provide sufficient validation? The designers themselves seemed to discount the standing of their radios and the eminent chroniclers of Art Deco have ignored them for 70 years so there needs to be a strong argument with supporting data.

If we are to present an impeccable case for radios to be accepted as icons of the Art Deco movement, I believe some questions need to be posed and that the answers are clear and compelling support for the thesis.

Do the radios presented stand distinct as a group?

The radios in this book constitute a small sub-group of radios manufactured in fifteen countries from the late 1920s to the early 1950s. The majority come from the USA, UK, Europe, and Australia in the 1930s and 1940s, but in some countries the effects of war and social upheaval delayed production of similar style radios by six to eight years (e.g., Czechoslovakia, Russia, Spain). The preponderance are tabletop radios and a majority are made from synthetic resins. The timeline of these radios almost exactly parallels the Art Deco period.

What is common in these radios that reflects the ethos of streamlining and Deco design elements?

- The use of new materials—plastic, glass, and chrome rather than wood;
- Mass production
- Produced cheaply, allowing broad consumer access
- A dedicated housing for the chassis, valves, dial, and speaker, which has smooth, rounded, or curved surfaces and aesthetic appeal
- The use of parallel lines, circles, squares
- The use of bright, contrasting colors
- A minimum of ornamentation
- A sense that the radio has been designed for the future, not as a link to the past

Are they individually and collectively of a standard that merits inclusion with traditional Deco objects?

Some, like the Sparton "Nocturne," are so special that they ooze the essence of Art Deco and Streamlining. Almost all are impressive individually, but as a collection they are vociferous team-members of the Art Deco squad, although each has a slightly different uniform. Today these radios are eye-candy with their elegant styling, beautiful colors, and rounded, glossy contours—all-in-all, quintessential Art Deco.

Interestingly, the fact that so many of these radios are so difficult to source today renders their mass-production somewhat irrelevant. The question of whether rarity confers greater merit on an item is an unresolved conundrum. However, some of these radios were only produced as samples and this imparts rarity immediately. As well, many were part of limited production runs (whether the model or the color) and given the fragility of the materials, the obsolescence of valve radios, and the length of time since manufacture, those remaining are now important examples. The fact that the majority were mass-produced, cheap domestic appliances should not preclude their being judged by the same design features as other Art Deco treasures.

Are there sufficient industrial designers listed to consider them important in the evolution of these radios?

We have statistical power in our samples. I have identified thirty-five designers in five countries who designed some eighty different models of radio for twenty-four different manufacturers.

Most of the designers (twenty-two) were in the USA (as would be expected) but the UK was well represented with ten designers. In the main, the output of these industrial designers is concentrated between 1930 and 1947, consistent with the introduction and growth of the Streamline movement.

Are some of the industrial designers included so important in the development of Streamlining that their radio designs cannot be ignored?

In the USA, names such as Raymond Loewy, Walter Dorwin Teague, Norman Bel Geddes, Harold van Doren, and John Vassos are mentioned in every monograph on Streamlining. Similarly, those designers responsible for radio design in the UK—Wells Coates, Serge Chermayeff, Mischa Black, and Jesse Collins—are synonymous with the modern design movement.

To be able to display multiple radios by all of these distinguished designers validates the radio as a utility embedded in the array of classic Streamlined objects of the time. There may be more famous items by each of the designers, but almost all of the radios are clear and timeless examples of their prodigious talent.

Is there a global spread to both the manufacture and design of these radios?

There are more than 300 radios in this book from 15 countries and involving in excess of 80 different manufacturers (see Gallery Index). The conception is in Europe, the birth in America, but the development is worldwide. This breadth and depth of dissemination speaks to the global nature of Art Deco and adds to the contention that these radios represent a cogent, comprehensive, and international illustration of the style.

Perhaps what is extraordinary is that not all the most beautiful radios were from the hands of these named designers, but it seems inescapable that these were all influenced by the efforts of the designers, and the benchmark of excellence and novelty they established. As well, there is ample evidence of the impetus of Streamlining in radio styling spreading right across the world. In Australia, for example, there were many stylish radios produced in the 1930s and 1940s by AWA and Astor and though the advertisements make claims such as "the radio was designed by an artist of distinction," it is not known whether these were independent commissions or in-house efforts by an individual or a design team. Bang and Olufsen, in Denmark during the late 1930s and early 1940s, had a design "team" but, at that time, rarely individually specified the members.

Talent is not restricted to famous names and the lack of identified designers should not stop us from appreciating clever and unique styling in a radio. The famous designers give us context and impetus, but it is clear that there were many skilled and pioneering people involved in the production of this group of radios.

What follows is a small selection of beautiful radios from across the world without a named designer. They illustrate the level of innovation and quality achieved in such diverse countries as Spain, Australia, USA, Mexico, and Denmark.

• The Artes A3 (c. 1947 Spain), a late entrant but so resonant of car shapes of the 1930s and 1940s).

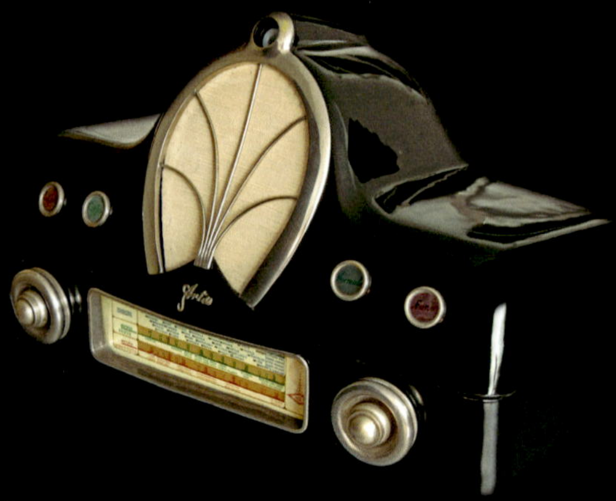

• The Australian AWA Radiolette (1936), with its robust architectural stepped sides, Chinese fretwork, and feet resembling lion's paws.

• The Fada Bullet (USA 1939), with its asymmetrical yet prototypical Streamline design plus mix and match colors for the grill, bezel, knobs, and handle.

• The EGM from Mexico (c. 1940), with its amazing curved chrome frame surrounding a circular radio with Aztec motif.

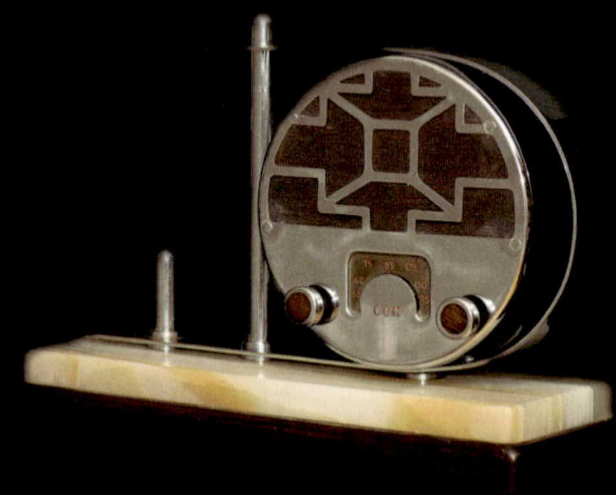

• The Bang & Olufsen Beolit 40 (1940), from the B&O design team in Denmark with its wonderful tapering and streamlined front.

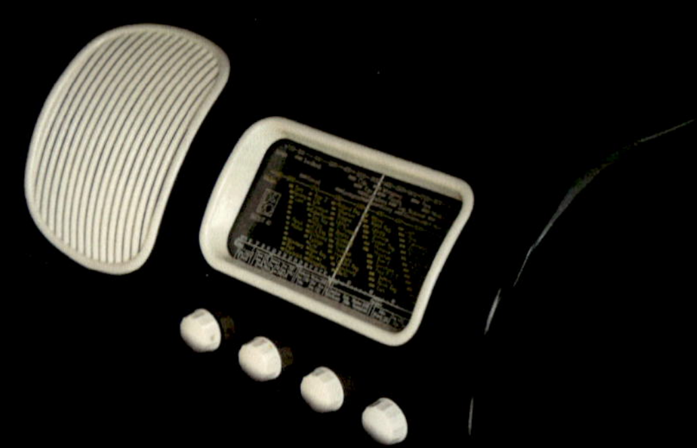

In all countries there was, over time and to a varying degree, the incorporation of local character into radio styling. America introduced streamlining to the radio world, opened the door to colored cabinets and with their mirrored and Catalin radios produced the icing on the Deco radio cake. The UK showed the world the possibilities of modern styling with plastic through Ekco's round radios and were among the first to jettison the old configurations for the modern shapes. The Ekco radios spread through Europe in the mid- to late 1930s but European radios themselves were stylistically stunted in the 1930s and 1940s as conflict prevailed. There is a somber and stolid tone to radio design, resulting in not only a limited variety of shapes but a limit to the coloring as well—to the muted colors of wood and brown or black Bakelite. However, there is something distinctive about French radios of the late 1940s and early 1950s, which incorporate an instantly recognizable Gallic flair and flourish, perhaps embellishing streamlining with a more decorative European approach.

Ekco Radios
Kaunas Lithuania
late 1930s

The Australians quickly absorbed the UK and US approaches and produced radios of quality and individuality. Most Australian radios were offered in multiple colors, perhaps more than in any other country. Astor Mickey and Football models of the 1940s in all their diverse colors reflect the experience and confidence of Astor to produce cabinets that were not copies of overseas radios but calculated to appeal to the Aussie consumer and home decor.

The adoption of modern styling and motifs in radio cabinetry across the world in the 1930s was not dictated by any commitment to a particular design model or theory, but more by a commercial implementation of a change in consumer culture and a general rejection of Victorian embellishment in favor of simple aerodynamic lines. In Europe in the 1930s, Art Deco sculpture and clocks often included classical features such as women and dogs, loosely based on Diana the huntress and her hounds. Streamlined styling for radio cabinets generally avoided this sort of ancient reference, but the wooden grill overlay in the Australian ESM mantel radio of 1934 is a beautiful, sleek adaptation of the hunting dogs. The same company fashioned the chrome escutcheon of the "Aristocrat" wooden console in 1935 (see below). This is more detailed in its treatment of the hunting dogs, clearly being Borzoi (Russian wolfhounds), yet still in the Deco style and in both radios the styling is more effective without any human adjunct. Interestingly, the representation of a woman and dog became very popular in advertising and fashion during the Art Deco period.

CHAPTER 7

INDUSTRIAL DESIGNERS & RADIO DESIGN

Harold van Doren

Raymond Loewy

Norman Bel Geddes

The role of industrial designers was central to the success of the Art Deco style radio, particularly those radios utilizing molded and cast resins. The 35 designers listed created at least 90 different radios working with some 20 different companies. More than fifty percent are in resin, the remainder are wood and mirrored radios. The bulk of the design momentum is in the 1930s, but the effect is seen for at least another decade. One of the features of these radios that spans the period 1930 to 1950 is that all the diverse strokes of the Art Deco movement, as well as its evolution, are visible. By 1950, there was a shift to modern plastics. The valve radios adopted a post-war "atomic" style before the advent of the transistor radios in the 1960s.

In this chapter, I will focus on the particular radios created by these designers, but it is the extended impact of their efforts on radio design in general that allows us to showcase in *DecoRadio*, in addition to those named designer examples, hundreds of beautiful radios that had no particular creator, yet exemplify all the best attributes of Art Deco styling.

Today, we all accept the importance of industrial design in the twentieth century as the cornerstone of a whole raft of innovative and trendsetting new products. For the industrial designer, the remit extends beyond form and function, being an applied science involving technology and economy. Additionally, there is the broader context of society as a whole, and, in satisfying and transforming the tastes of the consumer, industrial designers have changed our world.

The term "industrial design" enters the lexicon somewhere around 1920. There is a catalog of significant contributors to this field whose work in the 1930s, 1940s, and 1950s shaped the cities, the transport, and the homes of the modern world. Most of these men (industrial design appears to have been an almost exclusively male profession) are known for their buildings, their boats, planes, trains, cars, and furniture—a glorious array of design solutions— but are rarely lauded for their radio designs.

Radio, in the first instance, was a purely functional assemblage of technology—separate containers for amplifier, battery, and speaker, with little consideration for their presentation as a grouping. From the late 1920s, when the radio became a single unit, cabinet style was in the main dictated by furniture design. For the wooden floor console this followed the form of the phonograph (a box on four legs) until the mid-1930s when the box extended to the floor with small or no feet. There were also "highboys" and "lowboys," but eventually the cabinet morphed in the 1950s into a lower, wider radiogram that housed both the radio and record player.

The console retained Victorian and Edwardian styling well into the 1930s, but in time was replaced with stolid, modern features, including stepped shoulders and vertical elements such as fluting. Decorative approaches included mixes of veneers, inlays, imported marquetry, and beautiful wood finishes. Many wooden consoles could be viewed as boxes on legs begging to be used as firewood, but enthusiasts see them as the "Cadillacs" of radios and, in truth, there are some beautiful examples. Consoles with Deco styling such the RCA 9K10 from the USA (p. 168), the French Radialva TO-56 (p. 173), the Australian Astor "Minnie" external speaker (p. 126) and the glorious Pacific "Elite" from New Zealand (p. 138) rate a deserved mention. The Zenith 1000Z "Stratosphere" (P. 129) from the USA is a magnificent console with some Deco features but was primarily designed for performance and acoustics. The cabinet included walnut burl, Australian laurelwood, African satinwood, and Carpathian elm burl. Only 350 of the "Stratospheres" were made and the price, $350 in 1934, excluded all but the very wealthy. Interestingly, the popular use of motifs from ancient cultures in Art Deco is seen in the optional ceramic cat that fits in the recess on the front—a clear copy of the Egyptian feline deity.

The advent of the Depression caused most radio manufacturers to rethink their marketing strategy as profits fell. The tabletop (or midget radio as it was originally called) came into being in 1929 and had more scope for design variations than the console. It was not only a novel object in the home, but could be placed in other rooms than the lounge and serve other functions than as a family focus. Initially, it presented as a smaller variation of the console in cathedral or tombstone shape, perpetuating gothic styling using wood veneers. However, 1930 saw the first use of Bakelite as a material for radio cabinets, coinciding with industrial design developing as a legitimate and recognized profession. Actually, the Depression limited the work opportunities for most designers. In the USA, it was not till 1933 that industry (including radio) again commissioned designers such as Loewy, Van Doren, and Teague to find new ways to attract consumers. Thus, in 1933 the tabletop radio was proceeding down a number of pathways, some traditional but some very much reflecting new ideas on styling and design. In most countries from 1930 to 1933, old style was very dominant, but by the mid- to late 1930s modern styling was becoming the norm for the mantel radio, with the console more slowly embracing a modern external configuration. The popularity of the mantel, midget, or table radios is seen in the sales figures for radios in the USA in 1933. Of 3,800,000 radios sold, 2,250,000 were mantels, 850,000 were consoles, and 720,000 car radios. Some sixty percent of radios sold were mantels and three times as many mantels as consoles were sold in that year. It has been estimated that in 1933 half the radios in the world resided in the USA.

In the UK, Ekco, Murphy, and Kolster Brandes actually transferred to modern styling very quickly after 1931 with Ekco moving to Bakelite and Murphy and Kolster Brandes remaining with wood. In the USA, the popularity of radio and the advent of the smaller mantel was for some major manufacturers a sufficient increase in production and revenue to maintain the traditional materials and styling (wood cathedrals and tombstones) matching their wooden console range. In fact, most of the major companies (Philco, Zenith, GE, and RCA) moved to Deco styling in the mid-30s, but resisted producing Bakelite models till the very late 1930s. It was left to smaller innovative companies such as Air King, Colonial, and International Radio Corporation in the early 1930s to champion phenol- and urea-formaldehyde, while the FADA Company of New York initiated the use of Catalin in 1936. Some companies, such as Atwater Kent, never recovered from the Depression, but this might also be because they only made traditional wooden radios and simply did not address the changes in style, size, and material that attracted new consumers.

In Australia, the Amalgamated Wireless Company of Australia (AWA) was testing the waters in 1932 with two Bakelite models (C87, C104) exhibiting some Deco elements but set the standard for Australian radios in 1934 with the colored series of Empire State Radiolettes. By 1937, most of the other companies (Astor, HMV, Mullard, Airzone) were also offering phenol- and urea-formaldehyde models in multiple colors with modern styling.

Walter Dorwin Teague

John Vassos

Alexis Sakhnoffsky

Charles & Ray Eames

While not implying a complete cause and effect relationship between some very disparate events in the early 1930s, I would offer for consideration the interplay of the following:

- the expansion of radio as a communication medium
- the evolution of the tabletop radio
- the advent of Streamlining as a stylistic movement
- the rise of industrial design as profession
- the advent of new materials
- better methods of mass production
- cheaper domestic products
- mains electrification
- targeted consumer marketing
- a community desirous and capable of elective spending

They can all be seen as at least serendipitous and in some respects as consequential for radio as a necessary object in the home.

In the 1930s, radio was transformed by industrial designers imposing a new streamlining style ethic on this broadcasting vehicle. It became an object that consumers chose for their home not just on the basis of what they would hear, but how it would look. Radio became a visual as well as aural experience. The effect was that from a few ripples came a wave. Every company manufacturing radios was forced to modify their cabinet designs each year (whether or not there were internal technical advances) as choice and change became synonymous with retail marketing. The trend to modern styling became universal and almost all radios reflected the Art Deco aesthetic to some degree.

There is little in the literature yet the list of designers (both industrial and furniture) who conceived radios with Art Deco styling is extensive:

USA

Raymond Loewy
Colonial New World (1933)
Colonial 300 (1933)
RCA 568 (1933)
Emerson 511 "Moderne" (1947)
Emerson 520 (1947)
Emerson 547 (c1947)
Emerson 561 (1949)
Emerson 615 (1949)
Emerson 744 (1953)

Walter Dorwin Teague
Sparton "Nocturne" (1935)
Sparton 566 "Bluebird" (1936)
Sparton 557 "Sled" (1936)
Sparton 517B (1936)
Sparon 5470 (1937)
Sparton 409-GL (7-sided) (1938)
Sparton 059K (1938)
Sparton "Cloissone" (1939)
Truetone D2210 (1941)

Harold van Doren
Air King 37-39 (1933)
Air King 52-54 (1933)
Air King 66 (1935)

Norman Bel Geddes
Philco 370 "Lazyboy" (1931)
Majestic 651 (1939)
Emerson 400 "Patriot" (1941)
Emerson 544S (1948)

Clarence Karstadt
Silvertone (Rocket) 6110 (Sears) (1938/39)

John Vassos
RCA Portable Phonograph (1935)
RCA 6K10 (1936)
RCA 8T11 (1936)
RCA 9K10 (1936)
RCA 816K (1937)
RCA 96X (1938)
RCA A-20 (1939)
RCA "Little Nipper II" 9TX (1939)

J. Samson Spencer
Automatic Tom Thumb (1938 & 1939)

Frank Glover
Bendix 526C (1946)

Charles and Ray Eames
Zenith
Emerson 578 (1946)
Hoffman (1950)

Irving Rosenow
Majestic 776 "Lido" (1933)
Majestic 886 "Park Avenue" (1933)
Majestic 161 (1934)
Zenith 809 (1935)
Zenith 829 (1935)
Zenith 835 (1935)

Alexis Sakhnoffsky
Emerson BD-197 (1938)
Emerson AU-213 (1938)
Emerson AX-212 (1939)

Robert Budlong
Zenith 6d311 (1938)

Isamu Noguchi
Zenith "Radio Nurse" (1937)

George W. Walker
Detrola "Pee-Wee Midget" (1937)
Detrola "Super Pee-Wee" (1938)

John R. Morgan
Silvertone "Election" (1937)
Silvertone "Commentator" (c1940)

Barton T. Setchell
Setchell-Carlson 416 (1946)

Alexander Girard
Detrola (1946)

PHILCO DESIGNERS

Hollingsworth Pierce
Philco 511/512 1928 (metal table-top)

Matild Massaros
Philco 511/512 1928 (artist)

Albert Carl Mowitz
Philco Cabinet 1928

Edward L. Combs
Philco cabinet designs early 1930s

Clyde Shuler
Philco models 1932-33

Benjamin Nash
Philco 1934 (main cabinet designer)

UK

Wells Coates
Ekco AD65 (1934)
Ekco AC85 (1934)
Ekco AD36 (1935)
Ekco AD76 (1935)
Ekco AD75 (1940)
Ekco A22 (1945)

Jesse Collins
Ekco AC97 (1936)

Serge Chermayeff
Ekco AC64 (1933)
Ekco ADT95 (1933)
Ekco AC74 (1933)
Ekco EC74 (1933)
Ekco AD86 (1935)

J. K. White
Ekco RS2 (1931)
Ekco M23 (1931)
Ekco RS3 (1931)

Misha Black
Ekco UAW78 (1937)

Lawrence Griffen
KB BM20 (1947)
KB FB10 (1955)

Richard Drew Russell
Murphy A3 (1931)
Murphy A8 (1932)
Murphy A4 (1933)
Murphy A146CM (1940)

Eden Minns
Murphy

Betty Joel
KB Rejectostat 666 (1933)

A. F. Thwaites
Murphy A100 (1946)

ITALY

Livio Castiglioni
Phonola 573 (1939)

GERMANY

Walter Maria Kersting
Volksempfänger (1933)

NETHERLANDS

Louis C. Kalff
Philips PCJJ Loudspeaker (1927)
Philips 2113 "Permagnetic" Loudspeaker (1929)

DESIGNERS OF RADIOS

USA

Raymond Loewy

Colonial "New World" (1933)

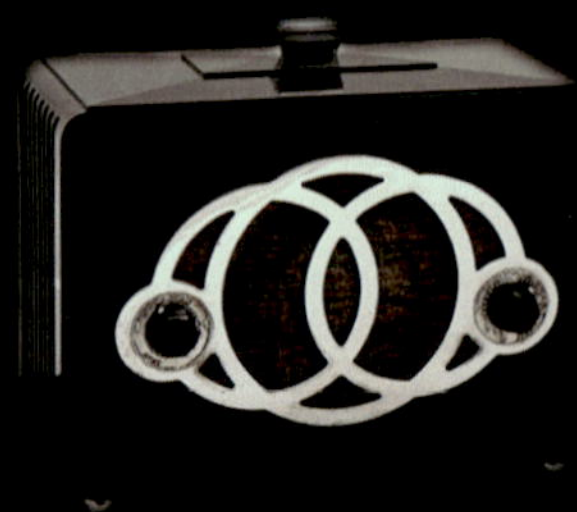
Colonial 300 (1933)

Emerson 511 "Moderne" (1947)

Emerson 547 (c. 1949)

Emerson 561 (1949)

Harold van Doren

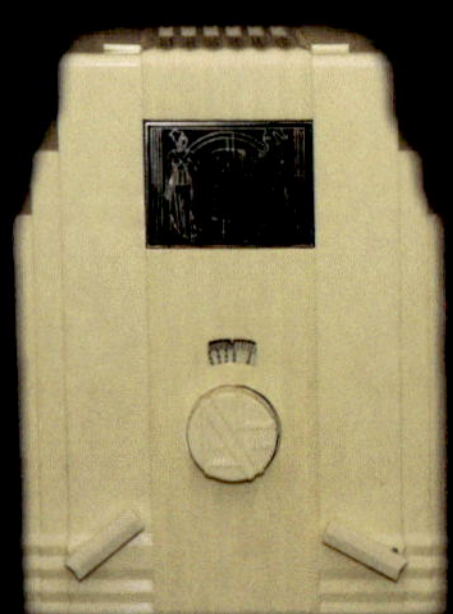
Air King 37–39 (1933)

Air King 52–54 (1933)

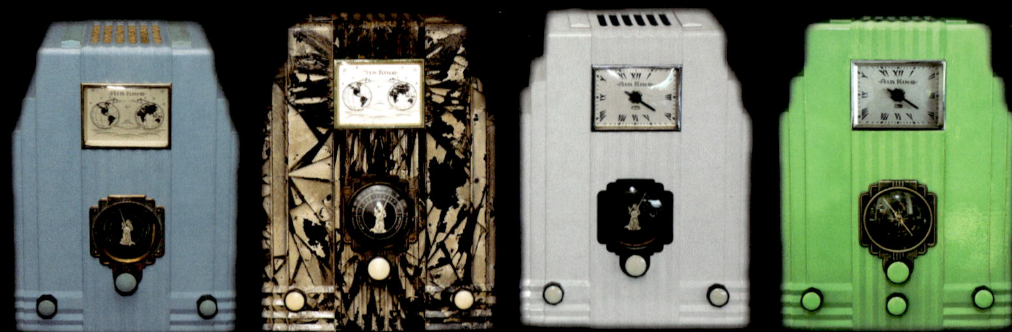
Air King 66 (1933–35)

Walter Dorwin Teague

Sparton "Nocturne" (1935)

Sparton "Bluebird" (1936)

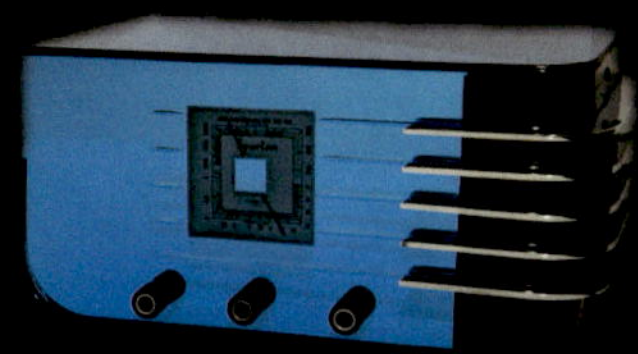

Sparton "Sled" (1936)

Sparton "Cloisonne" (1938)

Sparton (7-sided) (1938)

Sparton 517B (1936)

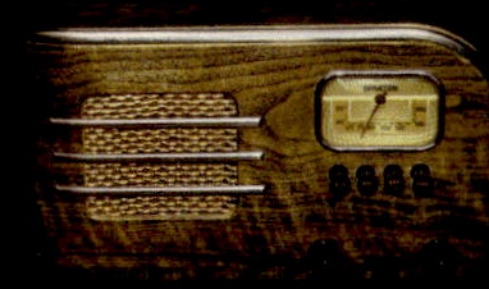

Sparton 5470 (1937)

Norman Bel Geddes

Philco 370 Lazyboy (1931)

Majestic 651 (1939)

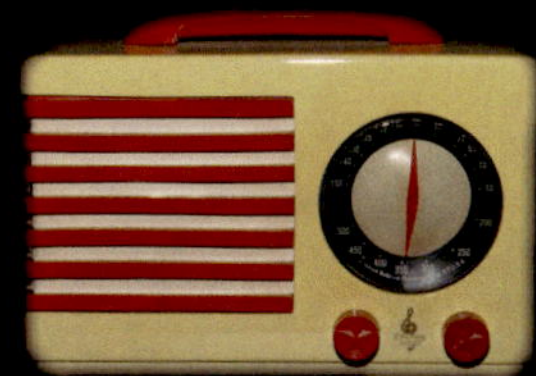

Emerson "Patriot" 400 (1940)

Emerson "Aristocrat" (1940)

Clarence Karstadt

Silvertone Rocket 6110 (1938)

George W. Walker

Detrola "Pee-Wee Midget" (1937)

Detrola "Super Pee-Wee" (1938)

John Vassos

RCA 6K10 (1936)

RCA BT11 (1936)

RCA 9K10 (1936)

RCA Phonograph (1935)

RCA "Little Nipper II" 9TX (1939)

RCA 96x-14 (1939)

J. Samson Spencer

Automatic Tom Thumb (1938)

Franklin Glover

Bendix 526C (1946)

Alexis Sakhnoffsky

Emerson BD-197 (Mae West) (1938)

Charles & Ray Eames

Emerson 578 (1946)

Robert Budlong

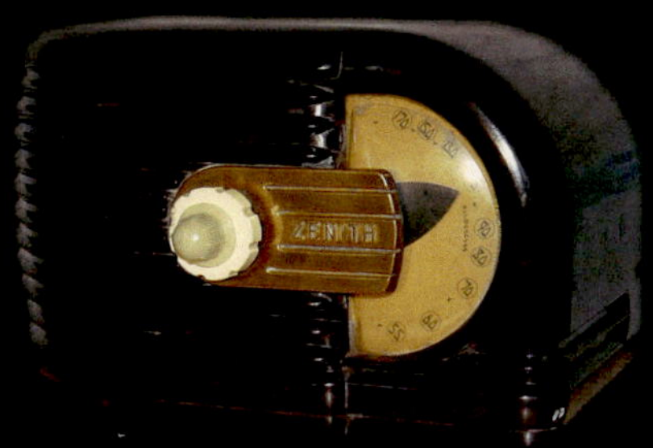

Zenith 6D311 (1938)

John R. Morgan

Silvertone "Election" (1936)

Silvertone "Commentator" (c1940)

Isamu Noguchi

Zenith "Radio Nurse" (1937)

UK

Wells Coates

Ekco AD65 (1934)

Ekco AC85 (1934)

Ekco AD36 (1935)

Ekco AD76 (1935)

Ekco AD75 (1940)

Ekco A22 (1945)

Jesse Collins

Ekco AC97 (1936)

Richard Drew Russell

Murphy A3 (1931)

Serge Chermayeff

Ekco AC64 (1933)

Ekco AC74 (1933)

Misha Black

Ekco AC86 (1935)

Ekco UAW78 (1937)

J. K. White

Ekco M23 (1931) & RS2 (1932)

Ekco RS3 (1932)

Lawrence Griffin

KB BM20 (1947) & LMT1840 (French version)

KB FB10 (1950)

ITALY

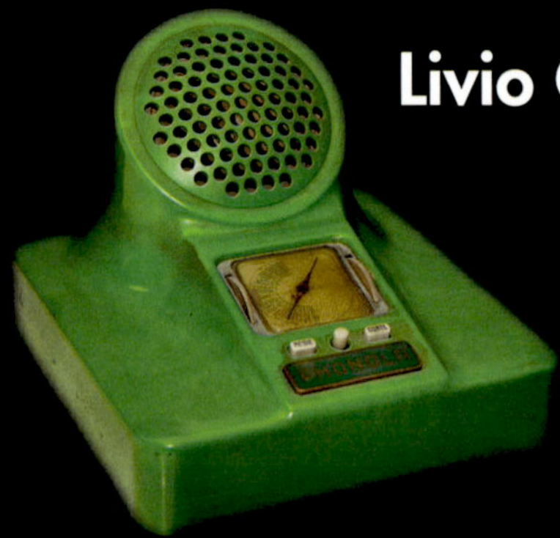

Livio Castiglioni

Phonola 573 (1939)

GERMANY

Walter Maria Kersting

Volksempfänger VE301 (1933)

NETHERLANDS

Louis Kalff

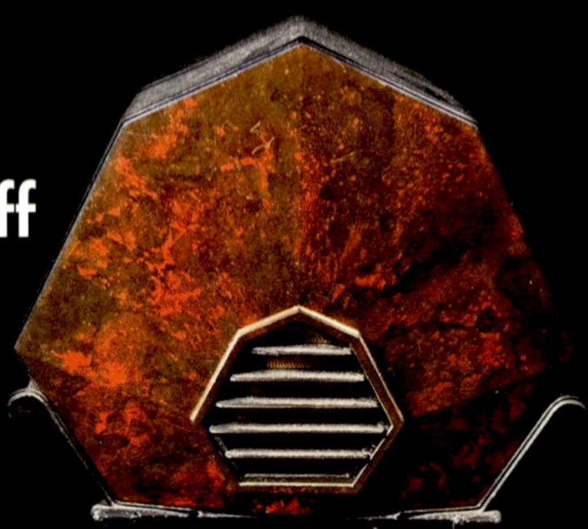

Philips 2113 "Permagnetic" loudspeaker (1929)

Philips PCJJ loudspeaker (1927)

These designers particularly embraced the new phenolic- and urea-formaldehyde resins because they expanded the scope of design and production while having commercial benefits. Others used novel materials, such as mirrored glass, aluminum, and chrome to showcase a new view of domestic appliances in the home of the future. Wooden radios constitute a minority, but the designs of Vassos and Sakhnoffsky clearly surmounted the limitations of the natural material and deserve inclusion. There are many attractive wood radios with Deco styling, but even with different veneers and variations in surface contour, they generally present a sense of consistent repetition, limited by uniformity in shape, size, and color. Plastic freed the designer to manipulate and become independent of these same parameters. The designer was restrained only by the physical constraints of the chassis, valves, and speaker. Each radio became a little or large work of art. Color alternatives also expanded the range of choice for the consumer within a particular model range.

Collectors of Catalin have made these radios among the most famous, cherished, and valuable, often without realizing their design heritage. It seems an almost global oversight that these radios rarely appear in the lists of credits for these designers. The radio companies were looking for another inexpensive addition to their line of wood and standard Bakelite radios. The unusual and the colored appealed to some consumers but they were not at the forefront of marketing or production strategy. For the designers, it was probably just an inconsequential but commercially necessary assignment in the post-Depression period that gained them little kudos or professional respect. Most of them moved on to bigger and better commissions, and the radios were never appreciated for their beauty until some 70 years later, and perhaps only today being seen together as a seminal influence in Art Deco and modern design. In the biographies and autobiographies of the most famous of these designers, their radios are generally conspicuously absent, or merely given the briefest of mentions.

It is only when we group these people together that a number of features become clear. These designers listed constitute the foremost and most famous. These men are the founders of their discipline and most are revered for their contributions. Almost all were involved in teaching and promoting industrial design throughout their careers, leaving a legacy of achievement that still resonates today.

Raymond Loewy has been called "the father of Streamlining," Norman Bel Geddes "the man who designed America" as well as "the man who streamlined the world," and Walter Dorwin Teague "the dean of industrial design." John Vassos, called "the quintessential modernist," chaired the boards of the American Society of Industrial Designers and the Industrial Designers Institute. Harold Van Doren was one of the founders and President of the Society of Industrial Designers. Misha Black, in the UK, founded the Artists International Association, was Professor of industrial design at the Royal College of Art in London and President of the International Council of Societies of Industrial Design. Wells Coates is regarded as one of the most important modernist architects in the UK.

As industrial design was a new field and modernist styling a new mode, most of our central characters came from other fields. In the USA, Loewy was a fashion illustrator, Geddes a stage designer, Teague an advertising illustrator, Vassos a commercial artist and graphic designer, and van Doren an art historian. In the UK Coates, Chermayeff, and Black were all architects. Although we are focusing on radio designs, we should at least mention some of their more famous commissions:

Loewy	Pennsylvania GG1 and S1 locomotives Sears-Roebuck Coldspot refrigerator Coca-Cola bottle and Shell logo Studebaker Avanti
Bel Geddes	Cocktail shakers Commemorative medallions Futurama Pavilion 1939 New York World's Fair
Teague	Kodak Brownie and Bantam cameras Polaroid desk lamp Steel legged Steinway piano Ford Building 1939 New York World's Fair

Wells Coates

Serge Chermayeff

J. K. White

Livio Castiglioni

Vassos	Peevy turnstile 1st TV for RCA in 1939
van Doren	Skippy racer scooter Plastic scales for Toledo Scale Wayne Gas Pump Maytag "Master Washer"
Coates	Designed the Isokon Building, Embassy Court in Brighton and 10 Palace Gate, Kensington

In the world of radio design, Harold van Doren's Air Kings of 1933 (pp. 116-119), Loewy's Colonial "New World" (Globe) of 1933 (pp. 122-122), and Wells Coates' round Ekco of 1934 (p. 139) individually and together herald a paradigm change for radio. This epochal time in both the USA and UK sees a revolution in both design and materials to present the growing popular medium to the public. Each of these models introduced modern styling and mass production, utilizing formaldehyde resins and color options. These radios seem almost alien when looked at in the context of the radio world of the time. In fact, they are even more extraordinary because, in the larger setting, they are indicative of the expansion of radio from hand-made furniture using traditional materials to mass-produced articles for the home and person, molded from colored synthetic materials. It is often stated that both the Air King and Colonial "New World" were designed for the World's Fair in Chicago 1933, which was to be the showcase for American design, but there is no evidence that they were exhibited.

The Air King was a collaboration between Harold Van Doren and John Gordon Rideout, the latter bringing engineering expertise to the project. This radio was not only innovative in that it utilized the new plastics to mimic the American skyscraper, but the choice of 11 colors and finishes was almost an assault on the featureless and conformist landscape of the competition and the taste of the American public. Today, we accept a range of colors and textures as standard for a commercial item, be it a car or a shirt or a table. But in 1933, these radios would have been a step into the new world of the future and in some ways a rejection of the standards of the past. We now recognize how successfully this radio satisfies the remit of the industrial designer: the adaptive use of a relatively new material in a novel product, mass-produced in colors never before offered and using a design motif mimicking the skyscraper, which was a symbol of the modern world. One of the features of the Air King that exemplifies the design and technical innovation was that the speaker was directed through the top of the radio, allowing greater design freedom for the front of the cabinet.

The Colonial "New World" radio designed by Raymond Loewy in 1933 is a masterpiece of inventiveness, artistry, and functionality. It can be seen as the symbolic embodiment of the impact of radio: radio waves circle the globe and the globe becomes the radio. Today, it is sometimes seen as a novelty, but this misses its innovative design elements and technical expertise at a time when all radios were boxes of some sort.

In another medium, Dorwin Teague also produced a game-changer with the use of mirrored glass and chrome in the "Nocturne" (p. 159) and "Bluebird" (p. 171) models of 1935/1936. These radios present a different approach to radio design, with a façade hiding the radio, which still resides in a wooden box. Despite the obvious dissonance between the form and quality of the front and back, the aesthetic impact of the front takes Streamlining to a new frontier. While only two dimensional, the combination of a circle of colored glass with horizontal and vertical chrome strips makes the "Nocturne" not just symbolic of the future, but a predictor of it. Teague's radio styling would not be out of place in today's high end audio-visual market. When a "Bluebird" comes on the market, there is often the story that the radio was a wedding gift for the sellers' grandparents and that it only came to light after they had passed on and the attic was cleaned out, the radio remaining unused as it had been too modern for the tastes of young marrieds in the mid-1930s.

The Sparton 7-sided (p. 202) continued the mirrored façade approach, but the 557 "Sled" of 1936 (pp. 174, 175) was a more refined, complete package. The wooden base was shaped like a sled with individual runners extending out from the cabinet. The front and top were in either blue or peach mirrored glass. The addition of chrome fins added to the impression of speed and motion. The wood was painted with a black lacquer, which created a rough crystalline finish.

Teague's tiny Sparton "Cloissone" of 1938 (pp. 242, 243) is probably the most cultured and evolved of all the Art Deco radios. With a Catalin cabinet and tenite (cellulose acetate) knobs, the front is enameled in one of four colors (red, blue, brown, and yellow) and decorated with chrome circles and horizontal lines. One of the circles contains the textured speaker cloth. It is the only radio with such a variety of materials and yet the effect is a seamless and subtle unity. It is perhaps a high point of the integration of form, color, and texture in the evolution of radio cabinetry, these criteria also forming the mantra of the industrial designer and underpinning much of Art Deco styling. For most collectors, this radio is always mentioned in reverent and wishful tones. Teague's radios for Sparton were acclaimed but did not sell in great numbers. All are relatively rare and consequently good quality original examples are sought after and expensive.

While our focus is on industrial designers, there were, in both the USA and UK in the early 1930s, furniture designers involved in radio styling. At this time, wooden radios, particularly consoles, were seen as pieces of furniture to be integrated into the living room décor. Both in the USA and the UK, companies such as Philco and Murphy hired furniture designers to revamp their console radios. Norman Bel Geddes was commissioned by Philco in 1930 to redesign their consoles, but was restricted in scope by the company and only the Lazyboy was produced as per Geddes designs. Geddes was trying to introduce functionality to furniture, but this seems to have been unacceptable to Philco at the time. Interestingly, for both Philco and Murphy, the long-standing involvement of furniture designers (whose medium was wood) rather than industrial designers (who were not wedded to any material) may have ossified the company's approach to cabinet design and been an impediment to extending their remit into the use of alternative and new materials, such as plastics. Murphy, for example, did not produce a Bakelite radio until 1940. Philco employed a long list of furniture designers in the 1930s: Edward Combs, Clyde Shuler, Benjamin Nash, and Edgar Haines. Although one of the largest radio manufacturers in the USA, Philco persisted with wood radios until adding some plastic models in the 1940s, starting with George Patterson, a Philco staff designer, creating the PT-89 in 1939, molded from Tenite.

In the UK, Richard Drew Russell was a furniture designer who had a long relationship with Murphy Radio starting in 1931 and designing over 40 wooden radios in ten years. Russell's A3 of 1931 was Murphy's first mains powered wood table model and also heralded the first use of veneered plywood in mass production in radio the UK. Most of Russell's radios were wooden mantels with some being converted to consoles. The A8 produced in 1932 was Murphy's first superhet receiver and the first cabinet in the UK with a landscape profile. Russell was followed by Eden Minns, who designed over 25 radios from 1937 to 1947. He also came with a background of furniture design and in many ways this seems to have left Murphy with a legacy of narrow styling, even when they eventually started designing and manufacturing Bakelite radios in the 1940s.

This is in sharp contrast to the Ekco Company, which vigorously explored new designs and materials for radios in the early 1930s, being one of the first companies in the world to truly embrace molded resins. The prescience of the owner, E. K. Cole, to gather around him some of England's best designers (Coates, Chermayeff, Collins, Black, and White) saw the firm rise to the forefront of UK radio manufacture in the 1930s. Without doubt, the five round models designed by Wells Coates between 1935 and 1945 are the highlight of British radio design. Each is a variation on the original theme of the AD65 of 1934, each with a different treatment of the central dial and speaker area. The last model, the A22 in 1945, is smaller, simpler in treatment, and perfectly symmetrical, a fitting end to an epochal series. Coates' series of round Ekco radios introduced a new geometric dimension to radio, emphasizing the versatility and potential of the new plastics and their ability to present continuous, smooth planes, which in some ways was the embodiment of Streamlining.

Not only did Ekco produce beautiful radios, but for a number of Coates and Chermayeff models they also designed matching stands, both in wood and tubular steel. The wooden stands are simple in principle, but this belies the fact that the radios slot seamlessly into the stands, which themselves have vertical elements matching those in the radios. The tubular steel stands are the essence of simplicity, and allow the radios to virtually exist without support.

The difference between furniture designers and industrial designers in the radio world of the 1930s highlights the particular issues that faced the latter group. Industrial designers were challenged by the technology surrounding mass-production of cheap radios with new and varied synthetic materials whereas furniture designers were still reflecting the more expensive craftsmanship of the artisan with traditional organic wood. Industrial designers were creating a small, portable domestic appliance capable of being utilized in many sites in the home by different consumer groups such as women and children, whereas furniture designers were restricted to larger, stationery pieces of furniture with a more restricted view of user dynamics. For furniture designers, each model was a single entity, but industrial designers worked with materials that allowed for variations in color and presentation.

A more particular example of design considerations was the fact that Bakelite was non-flammable, its first iterations being as electrical components. This allowed the cabinet to be designed in closer proximity to the hot valves, enabling a smaller and more portable appliance. Interestingly, while this worked well for Bakelite, it did not translate so well to Catalin, and the lighter colors often show evidence of tube-burn on the top of the cabinet, which, while not a major functional problem at the time, can today detract from both aesthetics and collectability.

One clear omission in design was the lack of ultraviolet inhibitors in Catalin. Thus, all white radios eventually turn to butterscotch (which is fortuitously quite appealing as a color), blues turn to green, and green turns to brown. They can be cut-back and polished to reveal the original color, but within a year or two the discoloring process is again visible. Some collectors believe that (like with old furniture) the patina of age should not be tampered with, while others try to restore the original status. I am personally conflicted between these two positions and have had a number of Catalins polished to bring out the original color and lustre. The fact that they subsequently regress and that repeated polishing thins the Catalin is probably a sufficient argument to tip me towards leaving the radios alone.

The move to streamlining in the early 1930s was not universally taken up in wood radios and for many models of mantel and console radios in the USA and Australia the transition was slow. In many respects the results were derivative, bland, and very similar in size and presentation. Sometimes the Art Deco effect was achieved with a distinctive treatment framing or decorating the speaker opening, either in polished metal or contrasting wood veneers. The Ingraham Company in the USA made high quality wood cabinets in the mid- to late 1930s for radio manufacturers such as Emerson and these are particularly collectible. Sakhnoffsky's Emerson (Mae West) is fascinating in that it is the only radio with an overtly sexual reference and this may have unfairly elevated its standing in the design pantheon. Seeing a clip of Mae West performing in her Madonna-type outfit quickly confirms the origins of the radio's outline (see p. 52).

The John Vassos designed wood RCA radios (8T11, 9K10, 6K10; p. 168)) are an interesting blend of wood and chrome. In each model, the cabinet has traditional but elegant proportions that would remain unremarkable but for the chrome tubular elements on the sides and base that wrap around the cabinet creating an intimate cradle for the wooden shell. The RCA 9K10, with its clever use of chrome mesh and tubular elements, converts the singularity of wood to a graceful blend of organic and inorganic materials. This overall augmentation of the wood console with complementary materials is rare and a reflection of the vision and skill of the designer. The 6K10 moves even further away from the wood legacy with its ebony lacquer finish that creates a striking contrast and complement to the chrome surrounds.

The RCA "Little Nipper II" 9TX (p. 235) is attributed to Vassos and is one of the few Catalin radios made by RCA. The Catalin version came in two colors: Brazilian green onyx with white knobs and white (now butterscotch) with dark red knobs. The design is simple and elegant, the face only relieved by narrow horizontal fins and the contrasting knobs. The same design was also available in Bakelite, plaskon (urea-formaldehyde), and wood. Interestingly, the cheapest version was Bakelite (walnut) at $9.95, plaskon (white) at $12.95, wood at $14.95 and the most expensive being the Catalin at $17.95. I have included Vassos's RCA Victor Special Model K, Portable Electric Phonograph of 1935 (p. 156) because it is one of the first uses of aluminum for a domestic appliance and because it is such a beautiful example of industrial streamlining.

Karstadt's Silvertone "Rocket" of 1938 (p. 207) is a wonderful example of machine-age design. Karstadt was a staff designer for Sears and it was released with the claim that it was "designed for modern people living in a modern world." Sears claimed it a sensational best seller and in 1939 also touted it as being designed to reflect the architectural splendor of the New York World's Fair.

Bel Geddes radio designs of the 1930s are less than spectacular, with the wooden Philco "Lazyboy" of 1931 excusable because of the intransigence of the Philco Company to major change at the time. The Bakelite Majestic 651 of 1939 is also unremarkable, but the hugely popular Emerson "Patriot" of 1940 (pp. 282-285) contains the spark of genius that in hindsight seems simple but reflects the talent of this innovative designer. Within an elegantly simple box framework, the shape of the radio and the contrasting colored elements channel the American flag at a time when war was imminent and patriotism a guaranteed marketing advantage. FADA competed with their tri-colored "All American" versions for both the Bullet (p. 289) and the L56 (p. 249) models, but these were not sold in huge numbers and now are valuable because of their rarity.

Toward the end of the Art Deco timescale, Loewy's Emerson 511 "Moderne" of 1947 (p. 322) seems to be an accretion of plastic boxes that attains no sense of coherence or fluidity. It is prettier in the dark with the dial globe on, but without Raymond Loewy's name would have no standing.

The post-war approach extended and exaggerated the angular aspects of the pre-war Streamline style using newer extruded plastics. But, at the same time, there were elements of Scandinavian minimalism that can be seen in the subtle and simple Eames wooden Emerson 578A of 1946 (p. 308). The clear plastic fascia and the black Bakelite knobs do not impinge on the clean lines of this small radio, which also reflects the muted post-war tone and ergonomics of the day.

It is interesting that the aforementioned industrial designers did not actually design many of the beautiful Catalin radios from 1936–1939. One reason might be that Catalin radios were a limited production run of an upbeat but costly alternative to the standard wood or brown Bakelite, constituting only a small proportion of radio production. Thus, for the larger, more established companies, this would not have perhaps warranted a design effort, and for the smaller companies that promoted Catalin radios, perhaps a designer was too expensive. This aside, the contribution of the designers is to have shown the possibilities and opportunities with the new materials.

The list of radio designers in the USA and UK is hopefully reasonably comprehensive, although I expect now that a list has been assembled, I will be confronted with a few omissions, which will cause me some sense of failure. More worrying is the lack of named designers in Europe, which might reflect language barriers in the research process. I have tried to include European examples of radios with modern styling, but the only named designers I could identify are Kalff in Holland, Kersting in Germany, and the Castiglioni brothers in Italy.

Walter Maria Kersting (1889–1970) was a German architect and industrial designer. From 1927 to 1932 he was Professor of Artistic and Technical Design at the Kölner Werkschulen. With his students, Kersting designed the Bakelite cabinet for the "Volksempfänger" model VE 301 radio receiver (p. 125), which was launched at the 1933 Berlin Radio Show. The simplicity of the design reinforces its functional limits in that the radio could only receive Nazi propaganda. Designed to be affordable for the masses, millions of units were sold and it became an iconic symbol of the voice of Nazism, but this was enforced popularity in the absence of any real local competition.

The Italian Phonola 573 of 1939/1940 (pp. 266-267) is one of the great examples of industrial design in radio and the first Bakelite radio made in Italy. People say it looks like a dictaphone or an intercom or even a set of bathroom scales, but it was designed to sit on the desk and reflect and integrate with the modern technology in the workplace. While I have attributed the radio to Livio Catiglioni, his younger brother Pier Giacomo and Luigi Caccia Dominioni are also credited with contributing to the design. The production of this radio is all the more remarkable seen in the context of the Fascist regime governing Italy at the time. Its amazing range of colors, including black, brown, burgundy, green, and red, as well as variety of knobs and dials make it a unique European accomplishment.

CHAPTER 8

THE PEOPLE'S RADIO

In the early part of the twentieth century, Germany was a focal point for many of the world's industrial and stylistic revolutions. Much of Modernism and Streamlining in the USA derived from Bauhaus functionalism. In 1931, the AEG company was at the forefront of machine age industrialization, making Bakelite cabinets for Ekco in England.

However, in Germany from the early 1930s to 1945, the emphasis was on indoctrination, combat, domination, persecution, and survival. The Germans were leaders in technology and ideas, but the Nazi ideology for an Aryan supremacy and appetite for war either ossified, reoriented, or destroyed any advantage. Many of the initiators of design and artistic movements were incarcerated or left Germany during this period. Radio in Germany during this period, both in purpose and in appearance, took a different pathway to the rest of the world.

By 1933, Hitler was in power and one of the first world leaders to understand and utilize the capacity of radio for political indoctrination. Goebbels said, *"I consider radio to be the most modern and the most crucial instrument for influencing the masses."* In the same year, the Volksempfänger VE301 (the people's radio) was introduced (p. 125), the Bakelite cabinet being the work of architect and industrial designer Walter Maria Kersting. The 301 model number commemorated the date of Hitler coming to power (i.e., 30th January 1933). The radio came in two sizes, costing approximately one to two weeks average wage. This radio was simple and unadorned except for either an eagle or a swastika above the central knob. It was clearly not meant for entertainment or enjoyment, but as a vehicle for the dissemination of Nazi cant.

The Aron Electrical Company diversified into radio in 1929, and with the growing anti-Semitism in Germany they named their radio Nora, this being the reverse of their surname. Nora led the world in 1930 with the Sonnenblume, the first Bakelite mantel radio (p. 106), and by 1933 Nora was the fourth largest radio company in Germany after Telefunken, Saba, and Mende. In an attempt to deflect attention away from the Jewish name, the core company was renamed Heliowatt, but in the end this made little difference and in 1935 the company was ostensibly sold to Siemens, but in fact was "aryanized" (a euphemism for "taken over by the Nazis"). The Aron family fled to the USA. In the grimmest of ironies, we see an advertisement for the Nora Volksempfänger in the mid-1930s (p. 125)—the Jewish company now fully integrated into the Nazi infrastructure yet retaining the original name.

All the German radio manufacturers became producers of radios designed for propaganda, much of their output being redirected for this purpose. They received grants to build cheaper receivers, with reduced capacity so that reception was limited to German controlled radio stations and could not

"All Germany hears the Führer with the Vollsempfänger" (German poster, 1936)

pick-up foreign or anti-Nazi broadcasts. It became a criminal act to listen to non-German radio stations and later in the war an executable offense in Germany and the occupied countries.

Between 1933 and 1939, more than seven million Volksempfänger radios were sold, being over forty percent of total radio production in Germany. Major manufacturers such as Telefunken, Siemens, Loewe, Lorenz, Mende, AEG, Saba, etc., all contributed to Volksempfänger production.

Yet within the sixty percent of commercial radio manufacture not dedicated to cheap radios for propaganda, there are clear hints of the emerging influence of modern styling in some of the more expensive wood and Bakelite radios produced in the mid- to late 1930s. In keeping with the German sense of seriousness and destiny, radio design at this time exhibits solid forms with extreme symmetry. Colors of German radios are muted (mainly black and brown) and there is little sense of the exuberance and innovation that we see in American and Australian radios of the same period.

The Telefunken 200 and Mende 148G from 1933 (see right and p. 124) show similar restrained lines within a stolid, traditional structural form. The imposing, near black Telefunken (nicknamed "the jelly mould") is arguably the second largest single piece of Bakelite moulding ever made for a radio cabinet and almost three times the size of the Mende. Both retain a cathedral shape with a vertical cascade from speaker to dial to knobs. Particularly with the Mende 148G, the curved, stepped shoulders and rounded top create in Bakelite what could not be achieved with wood, a harmony and fluidity of form and texture that gives the cabinet a sense of elemental completeness and streamlined smoothness.

The Lumophon WD210 of 1934 (see right and p. 137) is not a large radio by the standards of the Telefunken 200, but visually creates an impact that belies its size. It presents as a Bakelite edifice resembling a large public building or a concert stage. Whilst the bland Volksempfänger in its boxy orderliness says nothing of what is to be broadcast, the Lumophon calls the listener to attention with its framed centralized speaker. The brown pillars are buttresses topped with tiny stepped tiers. Surprising and rare is the use of a subtle, mottled grey colored molded resin (urea-formaldehyde), mimicking granite or marble, for the base and knobs, as well as the two square enclosures for the dial and speaker.

To round off the story of radio in Germany during the Nazi period, the 1936 Siemens 53GWL "Herr im Frack" (man in a dinner suit) provides a glimpse of the type of radio the middle and upper class would have been listening to during the war (see right and pp. 180, 181). This radio cost up to seven times the price of the Volksempfänger. An elegant piece of furniture concealing a radio, the black Bakelite cabinet with hinged doors sits elegantly on four white painted wooden balls. The four corner hinges are white Bakelite, as are the two round door handles. Otherwise, it is devoid of ornamentation and sits quietly, exhibiting the austerity of modern functional styling. Opening the doors reveals a superior radio with medium, long, and short wave bands and the dial showing stations in England, Russia, and Holland. Perhaps this is the radio that was used in the privacy of the home to listen to overseas broadcasts and was quickly closed if a knock on the front door suggested a visit from the Gestapo.

The idea of the radio as a vehicle for mass indoctrination was actually used before the Nazis by the fascists in Italy and the communists in Stalin's Russia. In 1931, the Mussolini government coined the slogan "the village must have the radio." Mussolini established a Ministry of Press and Propaganda, through which he exercised dictatorial powers over broadcasting. From 1933, the major radio manufacturers, including Philips, Phonola, CGE, and Telefunken, produced the standardized Radio Rurale that was supplied to communities, schools, and parishes right across Italy. The later models "Radio Balilla" (pp. 57, 195) and "Radio Roma" were intended to spread radio into the homes of all citizens. The "Rurale" and "Balilla" have impressive

Schaub 229 (Spitzkühler)

Tellefunken 200

Lumophon WD210

Siemens 53GWL (Herr im Frack)

"Traitor"
German poster (c. 1943)

Deco chrome motifs and lettering on the fronts of the small wooden cabinets, with both incorporating Mussolini's fascist axe-head.

The Philco Company in the UK released the Bakelite 444 "Peoples' Set" in 1936 (p. 182), but this was more about capturing the cheap end of the market with a low-cost, mass-produced item than having any political overtones. The English did produce a Utility radio (wartime civilian set) in 1944 to overcome the shortage of materials in the UK at that time, as well as the dearth of radios available. Some 175,000 of these cheap and bas sets were sold.

Like many countries (e.g., UK and Australia), Italy in 1931 promoted a local radio industry by imposing heavy duties and taxes on foreign radio and components. In most cases, this created the impetus for homegrown radio designs in each country, although there was clear influence from overseas models.

Radio Balilla (1937)

Italian children listening to Radio Rurale (c. 1934)

CHAPTER 9
THE AMERICAN PHENOMENON

While radio was a worldwide phenomenon, the uptake in the USA of the mantel radio in the early to mid-1930s was extraordinary. The rush to satisfy the householder at all societal levels meant that many wood and plastic radios of the time are both aesthetically and technically unremarkable and derivative. Even department stores and discount chains had their own brands of radio: Sears (Silvertone), Montgomery Ward (Airline), Gambles (Coronado), Delco (Belmont), W.T. Grant (Grantline), and Western Auto (Truetone). However, the American contribution to radio materials and design is special. Catalin, mirrors, plaskon, chrome, plus mix and match colors are either distinctly American innovations or reached new heights in the US context.

America championed small radios, exploring the uses of many forms of plastic and the broad appeal of color. Although the plaskon (urea-formaldehyde) Air King of 1933 opened a worldwide portal to both plastic and color for radio, it is the Catalin radio that is today seen by many collectors as the pinnacle of radio aesthetics. They are rare and coveted, with certain colors and models in good condition bringing prices in the tens of thousands of dollars. Major auction houses, such as Bonhams, have conducted a number of auctions of single owner radio collections in the last five years with the Catalins being the major drawcard. Today, the Deco radio has been elevated from a lowly collectible to the status of a valuable antique. It is likely that as the centenary of these radios approaches they will move up another notch in prominence and worth.

The introduction in 1933 of the Harold van Doren-designed plaskon Air King (pp. 116-119), with its architectural shape and palette of colors, is a defining moment in radio design. The Air King was not a single model in a number of colors. The core cabinet had a cut-out on the face that either fitted a map of the world, a clock, or an "Egyptian" scene. There were two models, one with three similar knobs (four for the export model) and one with a central, large, circular, Deco-style knob with two rod-like handles below. The latter seems ideally suited for the Egyptian insert and in many ways this combination has an elegance of vertical and horizontal elements that transcends the skyscraper motif, creating something almost like a ziggurat, a memorial, or a waterfall. It appears that the buyer could make a choice of insert (clock or map for the three or four knob versions) and thus many variations are seen. It should be noted that the 1920s had been a period of great interest in Egyptology with the discovery of Tutankhamun's tomb, and the ornamentation and imagery influenced many Art Deco motifs. Strangely, on close examination, the Egyptian insert seems to have some Grecian influence in the costumes of the two characters, but the American public would not have known or have been concerned about these minor details. Few radio models since have offered such an expanded range of colors (11 to 13), and this is truly remarkable for 1933. How many were made and sold is unknown, but by 1935 the Air Kings were also being marketed under the name of Auburn and Acratone, indicating that excess production had been sold off to secondary companies to sell as rebadged versions.

The Catalin era of radios is fleeting and uniquely American, extending from 1936 to 1941 and then briefly reprised after World War II from 1945 to 1947. Developed in 1927 by the Catalin Company, cast phenolic resin had no fillers and was therefore more translucent and able to be offered in many bright colors. Catalin was initially used for costume jewelry and fashion accessories as a cheap alternative to tortoiseshell, ebony, jade, and other organic materials. By 1936, it was possible to produce small radio cabinets and the FADA company introduced the small and simple cut-out box-shaped Catalin 5F60 (p. 177) with grills in different colors. This was the inauspicious beginning of a truly remarkable range of radios that from this simple unadorned box evolved over a few years into a kaleidoscope of shapes and colors. Catalin allowed the full expression of American Streamline design, offering another alternative to the more popular wood and basic brown Bakelite radios and a more translucent and vibrant color palette than the urea-formaldehyde cases. The 11 Catalin models of the FADA Company of New York are even more impressive when seen in the light of the company's financial struggles with bankruptcy and take-overs during the same period it was producing some of the most glorious Catalin radios of all time.

Catalin is more labor intensive and is less stable than molded phenol- or urea-formaldehyde and so was marketed as an expensive but vibrant alternative. The RCA Little Nipper 11 9TX (p. 235) is a good example in that it was available in four different materials. The cheapest version was Bakelite (walnut) at $9.95, plaskon (white) at $12.95, wood at $14.95, and the most expensive being the Catalin in two color alternatives at $17.95.

The size of Catalin radios was (at least pre-war) limited by its physical attributes and manufacturing difficulties, such as shrinkage, cracking, and finishing. Most Catalin radios are small, although a few models made post-war are much larger (Air King "Duchess" (p. 315), Bendix 526C (p. 318), and the RCA "tuna boat" [p. 309]). One fascinating feature of Catalin collections that distinguish them from Bakelite and wood radios is that there is a standard configuration for display as almost all Catalin radios were 7.5"–10" (19–25cm) wide, 5"–8" (13–20cm) high, and 5"–6" (13–15cm) deep. Another major issue with Catalin is that it contained no ultraviolet filters or inhibitors and all Catalins undergo surface oxidization, which turns white into butterscotch, greens to brown, and blues to green. This change is superficial and can be removed with polishing, although the color changes occur again over time. When assessing Catalin radios, the base will often indicate the true color, as it has been less exposed to light. Swirling of secondary colors is very common with Catalin and with this technique every cabinet has a unique pattern. Occasionally, this makes it possible to trace the history of a particular example through photos and auction catalogues.

In the USA, some 28 companies (approximately one third of those manufacturing radios) produced 46 different models of cast phenolic radios between 1936 and 1941 and some 14 models after World War II from 1945–1947. To put this in some sort of context, in 1938 (the second most productive year for Catalin radios), nine companies released 11 Catalin models while in the same year Philco (one of the largest US radio companies) alone released some 60 models out of approximately 400 models manufactured in the USA. This gives a rough figure for Catalins of some two to three percent of models produced in 1938, which probably would mean much less than one percent of actual numbers manufactured and confirms their relative scarcity at the time. A minor offset to this back-of-the-hand statistic is that a majority of radio models were wood and offered no variations, while the Catalins came in three to four colors for each model.

The main companies producing Catalin radios were Emerson and FADA, with important models from Motorola and Garod. Many companies only produced one or two Catalin models (Sparton, Air King, Kadette, Crosley, Arvin, and Addison).

Bendix 526C

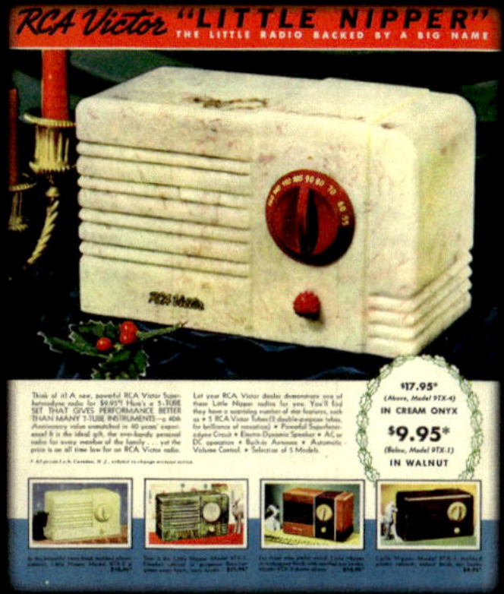

unpolished

1

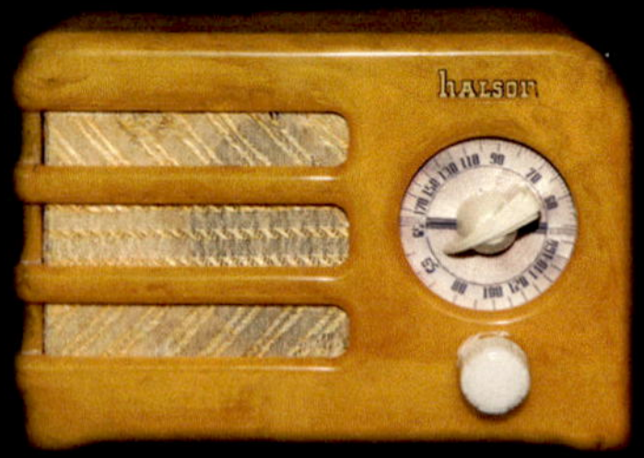
2

3

4

5

Most of the Catalin radios are beautifully presented in John Sideli's seminal 1990 book, *Classic Plastic Radios of the 1930s and 1940s,* and there is no intent to replicate the scholarship in this major reference. However, in the intervening years there have been other Catalin makes and models identified and photos of most of these are included in *DecoRadio*:

1 Detrola Split-grill 281 (1937), pp. 244, 245
2 Halson A5 "Dwarf" (1938), p. 205
3 Namco 601 (late 30s or 40s), p 303
4 Universal (c1939), p. 241
5 Sentinel 195U (1939), p. 247
6 GE L-622 (Jewel box) (1941), p. 280
7 Symphony (Split-grill) (1939), p.237
8 Stromberg Carlson (prototype post-1941), p. 292
Federal (n.d.)
Maguire (n.d)

One conundrum is the lack of named designers in the Catalin genre. At least one, if not both, the Automatic Tom Thumbs of 1938 and 1939 are by J. Sampson Spencer and the RCA Little Nipper II 9TX of 1939 is attributed to John Vassos. It might be argued that most of the companies involved in Catalin radios did not utilize designers at any time, but Emerson is an exception, having used Loewy and Geddes for Bakelite radios and Sakhnoffsky and Eames for wood radios in the period from 1938 to 1953. The Bel Geddes designed "Patriot" from Emerson is the only one of their eight Catalin models with a named designer.

As we have seen, Catalin radios were a colorful, but more expensive alternative to the cheaper standard wood, Bakelite, or plaskon radios, constituting a very small proportion of radio production. Though there are few named designers, looking at the evolution of Catalin cabinets in the FADA range, there was a clear process of experimentation and modification. It is hard not to see the hands of one or more designers in the accomplished L56 of 1939 (pp. 230, 231) and the distinctive Bullets (p. 291) and "Temples" (p. 297) released after 1940. Similarly with Emerson, the Catalin models show distinct changes between the tombstone AU-190 (p. 183) and the idiosyncratic "5 + 1" (p. 281) indicating neither a linear progression of cabinet form nor a standard approach to design.

Although generally small in size, the range of cabinet shapes plus the varied arrangement and color treatment of grills, handles, knobs, and dials is truly breathtaking. As an example, the grill in Catalin radios (which usually masks or highlights the speaker cloth) shows enormous variation and improvisation. They range from simple cutouts in the cabinet (Emerson AU190, Fada Temple) to patterns in the casting (RCA RC350, Addison A2, Automatic Tom Thumb, Motorola S grill 51x16). Many are a separate insert and almost all of these are in a contrasting color to the cabinet (Fada, Emerson). Most grills show a parallel linear arrangement with flat or fluted elements, both types often presented in a louvred arrangement. These can present as either horizontal (Fada L56) or vertical (Motorola 52, Arvin 532) and even split into two groups (Emerson EP375, Detrola Split Grill). Occasionally the grill will have circular elements, on its own as in the bezel on the Motorola Circle Grill (50XC) or interlocking interspersed with horizontal components as in the Garod 126. Most grills are a single color but that of the Emerson "Patriot" is distinctive with a two color pattern made from Tenite. The trend in Catalins was to disguise the speaker behind the plastic components. Using the speaker cloth as a feature is less common, but works well with the Motorola 50XC & 51X16 and the Emerson AU190. Probably the cleverest treatment combining both speaker and dial was achieved with the Kadette "Clockette" K25 of 1937, which uses a metal circular dial around the speaker grill and a large rotating altimeter style pointer.

Dials are generally on the right of the cabinet, but the Arvin defies tradition and places it on the left. Dials on Catalins are generally square or rectangular and occasionally round. Badging for Catalins was usually with a stick-on decal on the cabinet or integrated into the artwork on the dial. One great advantage of Catalin radios is that the vibrant colors and translucency of Catalin have an aesthetic appeal that augments and complements mere physical shape.

But to limit our review to Catalin is to miss some of the amazing American radios of the thirties. The Sparton company, through the work of Walter Dorwin Teague in 1935–38, offered four radios with chrome and colored mirror cabinets. These were the monumental "Nocturne," its baby brother "Bluebird," the "Sled," and a seven-sided version. All of these were available in blue and peach mirrored glass. As well, in 1939 they produced a small Catalin case with a chrome accented enameled front called the "Cloissone," available in blue, red, brown, and yellow. Even the rather humble Remler scotty in Bakelite was available with a colored mirror case as an optional extra. The radio slipped into the case with its white grill and knobs extending through the front.

Manufacturers such as Emerson, Detrola, Stewart Warner, and RCA produced a number of beautiful wood radios during the 1930s. John Vassos designed radios for RCA that all had aluminum tubular attachments surrounding the wood cabinet (RCA 6K10; RCA 8T11; RCA 9K10). His portable electric phonograph in aluminum is as modern today as it was revolutionary in 1935.

The "Radio Nurse" and "Guardian Ear" from Isamu Noguchi (p. 191) also bears a mention, although it is not a radio but a baby monitor. It was commissioned by the Zenith Company's Presidents in 1937 to monitor his daughter on a yacht and thought to have been in response to the Lindbergh baby kidnapping a few years earlier. The speaker is one of the few anthropomorphic designs in the radio world, having resonances of Japanese Kendo masks in a modern machine age style and material. It clearly represents the shape of a head, which can be further associated with a helmeted protector such as a hospital nurse.

Polished metal cases such as the Arvin 442 (1948) and 440 (1950) (pp. 327, 336) were only found in the USA, but in fact the painted versions of the same models were more popular. The rare chrome version of the Stewart Warner "Good Companion" (p. 264) is a round radio suspended from two chrome posts on a wooden base. It has a vitality and sense of movement that sets it apart from the ordinary radio normally solidly seated on a flat table, mantel, or floor.

6

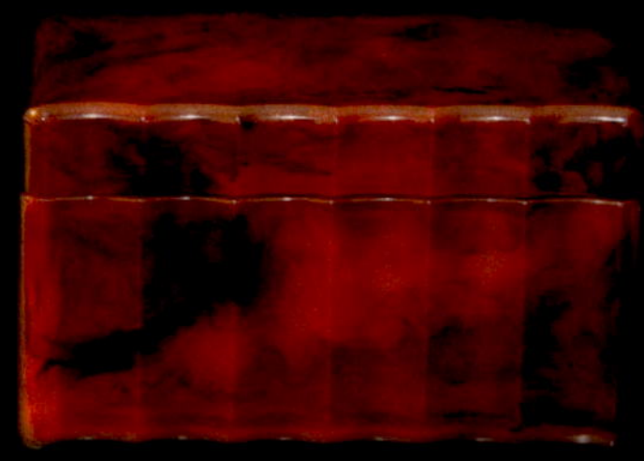

7

8

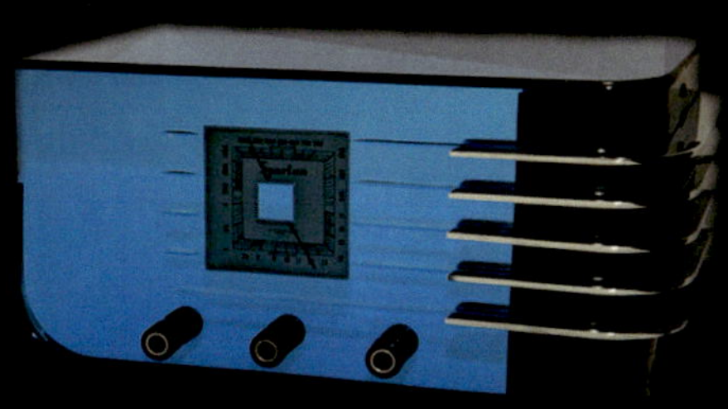

Sparton 557 (Sled)

RCA Phonograph

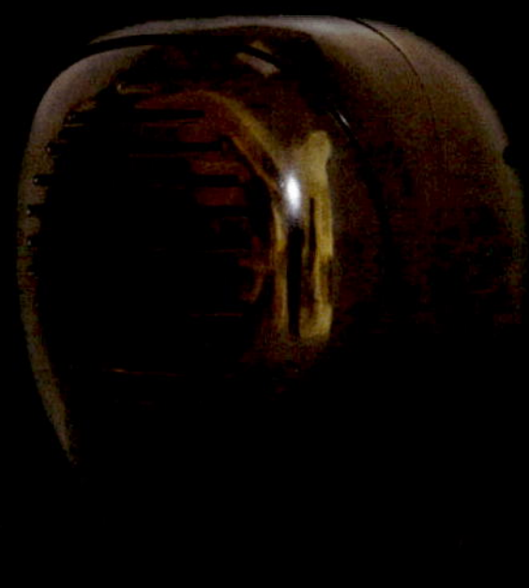

Zenith "Radio Nurse"

Arvin 442

Stewart Warner "Good Companion"

CHAPTER 10
COLORED RADIOS

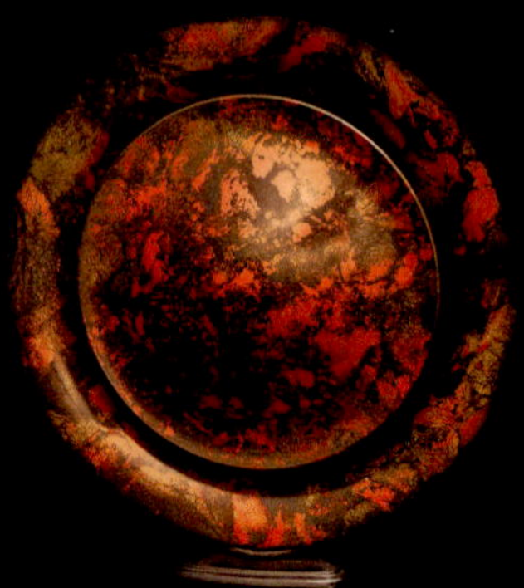

Part of the appeal of the Deco-style radio was the capacity for manufacturers to offer a variety of colors within a model range. This was particularly effective from a production and marketing strategy. Plastics, such as Bakelite and Catalin, appealed to a wider range of consumers and tastes than the wood radios, where the limits of the material tended to dramatically reduce the possible variations for each model.

The first evidence of color in radio appears to be in 1923 with the British Rexophone and Wyvern crystal sets (see left), which had a xylonite panel in a number of colors. Xylonite was an organic plastic made from cellulose. The French celluloid Minilux speakers of 1927 (see left) came in white and brown, with the former having scenes, motifs, and faces embossed or painted on the front convex surface.

The Philips company in the Netherlands made an innovative foray into color for radio components in 1927 with their futuristic Bakelite speakers (see left). Although these were a phenol-formaldehyde resin and limited to a palette of dark hues, at least seven different "colors" were offered. The plain brown and black were also augmented with marbleized reds and purples, plus a distinctive dappled green version. Gold flecking was also introduced in a random fashion in some examples. The design, colors, and performance ensured these speakers spread right across the world.

The Philco Company in America in 1928 offered painted receivers as part of their initial venture into radio production. The metal table model series 511–515 designed by Hollingsworth Pierce (see above) was basically a rectangular metal box sitting on an ornate cast iron stand (similar to the base of a Victorian sewing machine). The paint finishes were offered as Spanish Brown, Mandarin Red, Labrador Grey, Nile Green, and Impressionistic. To appeal to female buyers, artist Matild Massaros added hand-painted floral designs. Interestingly, Philco sold 96,000 radios in 1928, but in the 1929 model line-up only the Spanish Brown model was retained. Granted, this was the cheapest version in the series and this may have influenced sales, but America was clearly not yet ready for colored radios.

The advent of true color in radio cabinets (intrinsic to the material rather than paint) was initiated in 1931 in the USA and 1934 in the UK and Australia. All of these radios used urea-formaldehyde molded resins, the UK and Australian material being sourced from British Industrial Plastics. While in no way impacting on the sales of wood radios, colored plastic radios became within a few years a permanent fixture in the radio marketplace in the USA and Australia. Both of these countries were somewhat isolated (either by distance or topography) from the turbulence in Europe, which could be part of the reason why color did not appear there until after World War II. Color was attempted in England in 1934, but its proximity to Europe coupled with a propensity for wooden interiors probably helped to negate its impact. Emerging from the worst of the Depression, North America and Australia appear to have had a parallel path towards a broader consumer culture, with a lifestyle that embraced all the trappings of modernity, including objects in the domestic environment with modern styling and color variations. In both these continents, the ranch style, single level dwelling provided more space and rooms for the expanded use of multiple radios compared to the smaller European apartments and English terrace houses.

The extension of plastics as a material for radio cabinets in the early 1930s in Europe prompted the International Radio Corporation in the USA to release its tiny Bakelite Kadette "International" model in 1931. It came in brown, black, white, blue, and green as well as variegated marbleized versions that were quite striking. This was the first use of molded phenolic resin in the USA for radio cabinets and almost definitely made from Plaskon or Beetle (this being a urea-formaldehyde formulation rather than the standard dark phenol-formaldehyde Bakelite). The molding was produced by the Chicago Molded Products Company and while the shape of the cabinet was a simple box with cutouts for the speaker, the front had a traditional motif, hence the nickname "Gothic."

In 1932/33, the Kadette "Junior" was released and was the first pocket-sized Bakelite radio, being 6" (15cm) H x 4.5" (11cm) W x 2.5" (6cm) D. As well as standard dark brown, it came in light brown, a speckled brown, cream, and a bright red version (see right).

The Air King "skyscraper" series designed by Harold van Doren in 1933 is without doubt the transformative moment in radio design. While the shape and style was revolutionary using plastic, as well, there was a full array of 11–13 colors available, including black, brown, ivory, white, green, blue, lavender, and red. The rare red version has sold for over $50,000 at auction in the last few years. Additionally, there were finishes creating a flocked or crystallized surface texture. The black and white flocked example (p. 118) is a startling op-art installation and it is hard to believe it would be saleable in 1933, as well as being totally bizarre in the context of 1930s home decor. In the same year Raymond Loewy designed the Colonial "New World" radio (see right). This was a lavish experiment in design and production, with gold paint demarcating the countries on a black, brown, or ivory Bakelite globe, which itself housed the radio's chassis and valves. One very rare version of the "New World" radio has an actual paper globe applied to the Bakelite case, which can be dated to the early 1930s.

On the other side of the Atlantic in England, the Ekco Company had commissioned a number of well known architects to create radio cabinets using the new phenolic resins. From 1931 there is evidence from photos that the J. K. White designed models RS2 and M23 were offered in colors other than black and brown. Unfortunately, the images are black and white, so one can only surmise that these were white and green. In 1934, Wells Coates created the iconic AD65 (see right), a majestic mantel radio and the first round radio. Like the USA Air King of 1933, this is a pioneering triumph of industrial design and modern styling—a package for a new medium that had no precursor. As with the Air King, it did not just break the mold—it created a brand new one. Made from urea-formaldehyde, it was available in brown, black, green, and white. The round radio was a success with five different iterations being created over a ten year period. However, the colored versions were a failure from a sales point of view; today only four green are known to exist, and none of the white. Even the statuesque black AD65 with chrome knobs is quite a rare radio. This particular version was used by the Victoria and Albert Museum in London as one of their logos for the very successful Deco Exhibition of 2005.

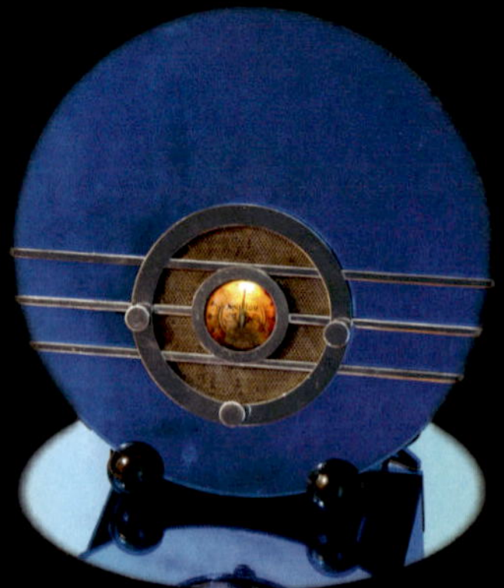

Images of Ekco product displays show colored varieties of the Wells Coates designed AC85 in 1934 and the subsequent round model AD36 from 1935. Only one white version of the AC85 is known to exist today. It may be surmised that this is one created for exhibition purposes only and none were produced commercially. From 1936 to 1945, almost every radio in the UK was brown and this default to dun could fairly be attributed to the frugality associated with wartime. In 1945, the A22 being the last of the Wells round series appeared in black and brown. There are a few colored A22's in circulation, which some have conjectured may have been sample models made for the "Britain Can Make It Exhibition" of 1946.

While colored radios were a small but consistent part of the radio market in the USA from the early 1930s, in the UK and Europe color was not really marketable until the post-war period and even into the early 1950s. However, one more country embraced color, and that was Australia. Without the benefit of named designers, in 1934 the AWA Company released the "Empire State" Radiolette (see left), which, though deriving its lineage from the USA's Air King of 1933, was a robust and independent solution in urea-formaldehyde. Available in brown, black, white, marbleized white, and green, it was well received, and though there are no production records to indicate the number of colored radios made, it is clear by the number still in circulation that the marbleized white was quite popular. The green AWA Radiolette is the most sought after Australian radio and can bring well over $20,000 for a pristine example. This is a fragile radio and prone to stress cracks and fading. Probably some 30 green examples still exist in reasonable condition and this indicates at least a moderate number were originally sold. Almost all green (pp. 166, 167) AWA Radiolettes are a jade color, but there are a couple of candy green examples that might indicate a different color altogether or simply a production variant. The pure white is actually quite scarce and one story is that this color was only produced for the New Zealand market that reflected a more staid and conservative British taste than Australia, which embraced the marbleized version. The rarest color is the speckled green, which is a one-off cabinet made for an exhibition and never had a chassis inserted (p. 143).

Australia's most unique contribution to radio design was the AWA Fisk Radiolette of 1935 nicknamed the "fret'n foot" (see left). This monumental tabletop Bakelite radio with massed sides, oriental fretwork, and feet shaped like lions' paws was one of the first experiments with contrasting colored elements. The brown model came with the fretwork and feet in white with brown marbleizing, the black either came with white fretwork and feet with black marbleizing or with vibrant green fretwork and feet. The rare green version had plain white elements. For some reason the marbleized white model had the same colored fret and feet. While each country developed its own style for radios, there was considerable poaching of elements from other countries. The Chinese style fretwork of the "fret'n foot" resembles very closely that in the International Kadette "Jewel" released in the USA in the same year. Perhaps this is a case of parallel evolution, but more than likely the Australians co-opted the pattern and scaled it up to the larger size radio.

In the USA in 1935, Walter Dorwin Teague designed a series of Sparton radios that even today could be considered avant-garde. Using round mirrored glass (blue and peach) and horizontal chrome elements, a round façade concealed a rather nondescript wooden case containing the chassis. Both the "Nocturne" and "Bluebird" (see left) are symbolic of Moderne and Streamline design. The "Nocturne," with its 48" diameter, was aimed for grand foyers, whereas its baby brother, the petite "Bluebird" (15" diameter) was, from a size point of view, suitable for the modern abode. Actually, these radios were so futuristic that they were famous, but few were sold. It is thought that 50–100 of the "Nocturnes" were made and now 10–20 are left, mainly in major museums and important collections. At least two "Nocturnes" have been sold for over $100,000. Teague also designed the Sparton 557 "Sled" (1936) in blue and peach mirrored glass. The same colors were available in the 1938 409-GL with its seven-sided beveled mirror glass front (see left).

In 1935, the fragile and now rare RadioGlo appeared with its glass panels and metal frame that lives up to its name when turned on and lit up (pp. 166-167).

The year 1936 saw the advent of Catalin being used for radio manufacture in the USA and heralded a glorious few years of riotous color in radio cabinets. The use of Catalin (a cast phenolic resin) extended dramatically the color palette available with the molded phenolics (e.g., Bakelite and Plaskon). Additionally, the ability to swirl secondary colors into the material prior to setting ensured an infinite array of results counterbalancing the conformity inherent in mass production.

Thus FADA in 1936 introduced the simple box-shaped Catalin 5F60 (see previous page left) with cut-outs for grill and knobs. The cabinet came in alabaster (which, because of the lack of UV inhibitors, are seen today as butterscotch) and onyx. The knobs matched the case, but the grill was offered in a range of colors (blue, red, green, alabaster, and onyx). For Fada and many other radio companies using Catalin, (e.g., Emerson, Motorola, Garod, Arvin, Symphony, and Detrola) the mix and match of cabinet, knobs, handle, and grill in different colors both highlighted and extended the color range for radios.

Catalin radios as a group are simply eye candy, with color complementing design in the best examples and compensating for blandness in others. It is hard to have a favorite as almost all are lovable, but the most impressive of the Catalin radios are the Fada Bullets with inset grills, the Motorola S and circle grills, the split-grill radios from Fada, Symphony, and Detrola, and Emerson's Little Miracle, 5 plus one and tombstone AU190.

One of the most popular American radios of all-time is the Norman Bel Geddes designed Emerson "Patriot" of 1939/41 (see right) using the American colors red, white, and blue. The Bakelite cabinet is either red, white, or blue and the other colors are reflected in the grill, handle, and knobs.

Probably one of the radios most favored by collectors is the Sparton Cloisonne of 1939, designed by Dorwin Teague (see right). It is a tiny radio with a metallic cloisonné front in either blue, red, yellow, or brown, trimmed with a lattice work of chrome elements using circles and parallel lines. The cabinet is usually white Catalin and the knobs are in white Bakelite. It is a true Deco masterpiece and highlights the mix of color, materials, and design.

Australia's most popular radio model was the Astor Mickey (molded phenolic) and over a ten year period from 1939 was offered in some 22 distinct colors and other variations, such as clouds, swirls, and mottling (see right and pp. 66, 326). Color variety seemed an important marketing approach in Australia with many radios in the 1930s and 1940s produced in six to ten different colors.

In England, the first indication of colored radios being sold in any numbers appears in the late 1940s with Kolster Brandes BM20 "Dalek" (see right and p. 323)and in the 1950s with the FB10 "Toaster" models.

In France, a limited range of colored Bakelite radios appeared post-war in 1949. These include a French version of the KB BM20 in various colors, the "Sonorette" in green, white, brown, and burgundy, and the Radialva Super Boum in mixes of colors, such as a white case with green grill and knobs (pp. 66, 306, 307), burgundy case with red grill and knobs, and brown case with white grill and knobs. The rest of Europe remained wedded to wood and dark Bakelite until the mid-1950s when modern plastics and then transistors took over.

The Russian Svezda (Red Star, pp. 66, 345) is a version of the French Excelsior SNR 52. Produced in the 1950s, it is predictably red with a star in the center that lights up when the radio is turned on. Here color has political overtones but without the malevolent propagandist intent of the understated German Volksempfänger of 1933. A late piece of Deco styling, the gargantuan tabletop Zvezda unambiguously and yet inoffensively uses color for purposes other than mere decoration.

The Italian Ducati "Paniere" of 1940 (pp. 66, 195) is one of the few wooden radios in the world that can be described as colorful. The clever use of green Bakelite knobs, green print in the dial, and a green matted speaker cloth add contrast to the light wood veneer and give this unusual shaped radio a commanding presence.

ASTOR MICKEY
CROSLEY
Motorola

CHAPTER 11
BRANDING

Today we would all recognize branding as a fundamental part of the marketing of radios. The name of the radio in full sight of the user, imprinted on their memory with every use, seems a simple and effective way of reinforcing brand recognition and longer-term allegiance to the company's products. If the radio is a good example of the commercial world of the 1930s, then it was surprisingly less sophisticated than today. While we see many strategies for producing attractive and profitable products in the 1930s, branding was not as pervasive or as overt as it is today. Not only were there diverse methods of branding that can be associated with the cabinet materials, but trends were also country specific. Many companies varied their branding approaches with different models. Sparton, for example, has the logo included with the dial on the "Bluebird," "Nocturne," and "Sled," as part of the plastic molding on the "Easy-Goer," a chrome badge on the "Cloissone," and etched into the mirrored glass on the seven-sided 409GL. What is more surprising are the number of instances where branding on the front of radios was completely absent.

The approaches to branding include the following:

- Metal or plastic badges
- Incorporated in the plastic molding
- Integrated in the escutcheon
- Etched in the station dial
- Decals

Wood radios occasionally have brand names on badges or within the escutcheon, but once large dials became common, the manufacturer's name was generally placed adjacent to the radio stations. When the radio was turned on, the illuminated dial highlighted the name of the radio as well as the stations.

Many European Bakelite moldings in the early 1930s had the makers name included on the cabinet or the grill, either recessed or in relief (Nora, Sonora, Loewe, Siemens). Some makers (such as Philips and Telefunken) had their logo as part of the Bakelite escutcheon surrounding the dial. In the latter case, Philips only displayed their logo, probably assuming that this was easily recognized and automatically associated with the company name.

Occasionally, the name in the molding was highlighted with paintwork (Genalex, Crosley, Majestic).

In Australia, every radio was branded either as part of the Bakelite molding, with metal or plastic badges on both wood and Bakelite radios, or with the company name etched on the glass dial.

Radio manufacturers in the UK seemed to be either reticent about emblazoning the name on the radio or confident of the public's recognition of their products. The Ekco round Bakelite series have no name at all other than on the cardboard backplate. The Ekco

AC86 has a tiny black and white plastic Ekco badge on the front, but the monumental AC97 (Robby the Robot) has no indication of maker or model. This trend also extended to Bush, whose DAC30 has only a little tree on the dial to represent the manufacturer. Similarly Kolster Brandes KBBM20 (Dalek) and KB FB10 (Toaster) are unnamed.

In America, the van Doren designed Air King plaskon series of 1933 to 1935 had nothing on the cabinet to distinguish the maker. On the three knob version, the name was only on the globe or clock insert (not even on the dial), while on the waterfall version with the Egyptian insert it was molded into the central knob. On the later wood Air King 355, the name Air King can be seen on the dial. The Loewy designed Colonial "New World" and 300 from 1933 have no indication of the manufacturer's name or model designation.

Another quite different trend developed in the USA in the mid-1930s, which was the use of cheap water-based stick-on decals, seen in some Bakelite and wood models but very prominent in the Catalins. Catalin radios have the maker's name either integrated into the tuning dial, which is usually covered by a celluloid window (Motorola, Arvin, Detrola), or with a decal on the front face of the cabinet (Emerson, Kadette, Garod, Symphony, Lafayette, Automatic Tom Thumb, Sonora). This has always seemed to me a rather pedestrian finish to a beautiful product. But one might argue that the name integrated into the casting or an added badge would disturb the smooth lines and curves of the Catalin radio and this simple, flat result is an homage to Streamlining. However, radios were a product made for profit and commercial self-interest would always outweigh artistic goals. It is more likely the cabinets were outsourced and the most frugal and generic approach taken to their production. Additionally, there may have been the potential for rebadging or remaindering of cabinets after the model run was exhausted. This occurred with Air King being sold as Acratone and Auburn in 1935 with a different name in the globe insert. In the end, it may simply be that branding was not seen as the important marketing tool that it is today. The rare Catalin Symphony split-grill is an exception, with a metal badge on the front of the cabinet. Another exception to the name being on the dial or a decal is seen with the GE Jewel Box, which has a metal badge on the internal surface of the radio that is not seen when the lid is closed and thus does not spoil the elegant fluted lines of the Catalin surface.

One of the effects of re-polishing Catalin radios is the loss of the original decal, which, even though damaged, can add a touch of patina to a 75-year-old radio. However, these days one can purchase a page of reproduction decals that will include most of the well-known brands. Placed in water for a few seconds and then slid onto the Catalin face seems almost too easy to be permanent, but effectively restores the branding to close to original showroom condition.

Metal radios generally had the brand on the dial (Stewart Warner) or on the plastic tuning knob (Arvin). An exception is the chrome EGM from Mexico, which has the initials in relief as part of the metalwork below the dial and highlighted with red paint.

An interesting story regarding branding relates to the Pye Company in the UK. In the 1930s, Pye radios had a grill design of a stylized sun with rays, which became the company's logo. In 1947, after the war, they released the portable M78F (pp. 328-329) with an acrylic case and using a Deco version of their emblem. Unfortunately, the similarity to the rising sun on the Japanese flag caused an intense adverse publich reaction. Pye recalled all unsold radios and destroyed them. Those few remaining are the ones that were sold. They never used the design again.

One of the most overt examples of branding is seen with the Ducati RR3404 of 1940, the company better known for its motorcycles. Nicknamed "Paniere" because it resembles a basket or hamper, the Ducati name is emblazoned over the speaker opening in wood lettering and takes up perhaps twenty-five percent of the surface area of the front of the radio. Suprisingly, the effect is impressive, even more so given that this radio was made during the war in fascist Italy.

A selection of decals

CHAPTER 12

KNOBS & ESCUTCHEONS

From inception, the radio has had knobs that allow for tuning, altering the frequency, the volume, and for turning the device on and off (the last function actually not an original feature on many early radios). Knobs came in wood, plastic, and metal, as well as combinations. Knobs serve both as a functional interface for the user as well as a decorative element. What is remarkable is that almost every model of radio has a uniquely shaped knob. This necessary accessory is generally subtle and, in the case of wood radios, never an aesthetic feature; but, in the colored radios, the knob was often used as a complementary counterpoint to the shade of the cabinet.

Knobs are rarely a major feature, but one example is the Spanish Artes, where the metal knobs have the same impact as headlights on the front of a car. Some colored radios came with matching colored knobs. Probably the best example is the Australian louvred Astor Mickey (urea-formaldehyde), which sports over 20 different colors and has at least 10 with matching colored knobs, the remainder using white or brown. The American Catalin radios generally used knobs of starkly different color to the case and usually matching the grill and handle.

All the early plastic radios have plastic knobs with a rare variation being the chrome insert in the black versions of the Coates designed Ekco AD65 and AC86. One other exception is the Loewy designed Bakelite Colonial "New World" where the knobs are gold-plated metal as part of the equator ring surrounding the Bakelite globe. Most Catalin radios have Catalin knobs, but Emerson (for example) used urea-formaldehyde knobs on many of its models (Little Miracle, Big Miracle, 5+1).

AIR
KING

Wood radios all originally had wooden knobs, but many consoles and mantel models in the 1930s and 1940s began to provide cheaper brown Bakelite knobs (Radialva console, Stewart Warner "Spade"). Small metal radios generally have plastic knobs (Arvin), but there are some unusual arrangements: the painted Spanish Artes has all metal knobs, the chrome Stewart Warner "Good Companion" has plain wooden knobs, while the chrome Mexican EGM has chrome knobs surrounding a wooden center. A small number of radios included plastic push buttons for tuning (e.g., Bakelite Silvertone Rocket, Catalin Sentinel, and wooden Stewart Warner "Spade").

Most radio knobs are either retained by a side screw or by frictional fit using a metal sleeve. Plastic knobs were generally Bakelite or urea-formaldehyde, but were also made in Tenite (a cellulose-based thermoplastic) and Lucite (a transparent acrylic).

Before plastics were used for radio cabinets, there was a twenty year period of Bakelite being used for small electrical components, such as plugs and switches. While cabinets were relatively costly and complicated, knobs were much easier to produce and with the introduction of urea-formaldehyde and Catalin we see an almost infinite variety of shapes and colors. Some knobs are just simple and circular, but most are edged, fluted, or faceted. These irregularities make sense for ease of gripping by fingers.

Today, one can buy—or have made—reproductions of virtually any wooden or plastic knob. Actually, because of the ease of making plastic knobs from a mold, many dealers offer plastic versions of what were originally wood knobs. Ideally, original knobs, even with slight damage, are preferable to new ones. However, a radio with a good cabinet is not radically devalued with the addition of reproduction knobs. Nevertheless it is considered good form to divulge information regarding parts that are either not original or repaired in any description of a radio to be sold. In the USA, a few people have purchased old stock Catalin rods (originally intended for jewelry and small kitchen items) to make new knobs for Catalin radios that are almost indistinguishable from the originals.

The escutcheon is a decorative element that frames the dial of a radio, usually in cast metal, chrome, or brass, but occasionally in plastic. It is most often seen in wood consoles, less commonly in mantel radios, and rarely in the plastic models. It is worth including here as there are many instances where the Art Deco styling of the escutcheon can lift the aesthetics of a radio out of the commonplace.

The Teague-designed Sparton "Bluebird" and "Nocturne" both have elegant circles and lines of chrome work surrounding the dial and highlighting the mirrored glass. The Air King 66, designed by van Doren, is unusual in plastic radios, having a metal escutcheon that not only houses the dial but also one of the knobs. This is effective against the colored plastic cabinet, but with the additional clock or globe insert situated above the escutcheon, it adds a bit of fussiness to the uncluttered uniformity of the cabinet's simple central vertical flutes. This is offset by the lack of a speaker enclosure on the front of the radio.

The Philips Arbolite series from the early 1930s all have black Bakelite escutcheons that have a radiating floral motif, in some cases extending beyond the dial to border the speaker enclosure. These escutcheons generally include the Philips logo and add a clear modern influence and contrast to the brown and red mottled Arbolite cabinet.

MEDIUM WAVE KILOCYCLES
LONG WAVE KILOCYCLES
BEALE
ATLAS
KILOCYCLES
MEGACYCLES
Sparton
BROADCAST
STEWART-WARNER
DETROLA
MEGACYCLES
FOREIGN SHORTWAVE
GRANDES O.
PETITES O.
LANGE G.

CHAPTER 13
COLLECTORS & COLLECTING

Collectors are generally normal people with a mild version of "Manic Acquisitional Disorder" (MAD). It seems to affect more men than women, is prevalent in all socio-economic groups, and symptoms can occur at any age. It is incurable and no medications seem to modify the compulsion to purchase. Some have postulated that the syndrome is caused by a microbe (the collector bug), but as yet it has not been isolated.

Another version of this syndrome is where the collector hides objects from significant others and then pretends they were bought sometime earlier. Also the cost is generally severely understated. This is called "Surreptitious Acquisitional Disorder" (SAD).

What makes someone a collector? At what point does a group of objects become a collection? What's more—who cares? For those who find this sort of navel-gazing opaque and self-indulgent, let your eyes glaze over the words as if they were just indiscriminate tiles on a Scrabble® board, shake your head in bemusement, and find the pages with beautiful pictures of radios.

I know that response intimately. I am the only one in my extended family who acquires with passion. The family does not ridicule me, but they do not understand me. When I mention to people that I collect old radios from the 1930s and 1940s, I engender a range of responses on a continuum: from perplexed tolerance to patronizing indifference—a sense of "he's crazy but, hey, it's harmless" from family and friends; "they're colorful but hardly worthy" from the more upmarket antique and art collectors; and "why bother with what was the cheap end of radio technology" from the majority of radio enthusiasts. I concede that many people think this sort of activity is a waste of money, of space, and of time.

Things of course are changing as exposure via television, radio, exhibitions, books, and auctions has bolstered the status of Bakelite and Catalin radios and reduced the disdain of the naysayers to a rictal smirk. Obviously, it's hard to argue with publicity and success, particularly if values are escalating. If you are lucky enough to choose to collect something that you love (whatever its initial status in the hierarchy of possessions) and the items become more valuable and more generally respected, people admire your prescience even if they never appreciate the "stuff" you have, much less understand the compulsion to build a collection. Add to this the general notion of anything made of "plastic" as cheap, insubstantial, and gaudy, and it is easy to see why people in general (and many radio enthusiasts in particular) dismiss these mass-produced artifacts as unworthy of being collected, much less revered.

But what, can we ask, is a collection and how many does one need to have a collection? Obviously, by any definition of the term, there needs to be a minimum number (perhaps five to ten) of the same sort of item to qualify for the label; after that, it's up to the collector to mark out the depth and breadth of their collecting field. As my collection grew from 50 to 100 to 250, there were consistent responses of "where the hell do you put them" and "well, there can't be many left you don't have." Actually there are no guidelines for collecting and the scope can be limited to instances of one particular creator (e.g., artist, sculptor, author, or company), a distinction within a collecting area (e.g., plastic radios, tin robots, Cubist artists), or expanded to a major field (e.g., contemporary art, cartography, photography). Less common today is the all-inclusive approach, which straddles many themes (e.g., cabinet of curiosities, *objets d'arts*).

What does it mean to be a collector? Freud thought collecting was a type of pathology based on poor toilet training, whereas Jung thought of collecting as a sort of archetypal drive, such as early humans gathering berries. Granted, collecting can mimic the lack of ultimate fulfillment seen in extravagant and compulsive consumer purchasing, but one would like to think there was something a little more virtuous in the assembling of antiques compared to the stockpiling of shoes.

First, let us distinguish collecting from hoarding. Someone with a barn full of old, unrestored, broken radios collects but is not a collector. People who have an odd piece of china, furniture, or artwork may prize these items, but that does not make them serious collectors. In my view, collectors are people who have defined the boundaries of their field, know their subject to the extent of being an amateur authority, and build their collection with purpose and conviction, always striving to procure the best examples available.

A true collector can be an expert in their field. There is enormous pleasure in researching the history of a piece, its design and manufacture, its place in the timeline and hierarchy of a particular genre, and the social and cultural context that it informs. Often the museums and galleries that purport to have radio collections have a few uninspiring examples, and those are usually incorrectly described and badly displayed. Radio seems to have never been appreciated in its own right and is often a very ephemeral bit-player in museum sections devoted to twentieth century design, science, or communication. Institutions, curators, and academics remain detached, showing little interest in improving their radio collections or broadening their knowledge base. It seems a shame that little or no advantage is taken of the goodwill of collectors who would happily provide radios for exhibitions and who embody an extraordinary reservoir of knowledge about all aspects of the subject.

People collect in many different ways and perhaps this muddies the waters somewhat. There are many accumulators and hoarders whose abodes and lives are crammed with items of dubious significance and value. Perhaps this is where the term "collecting" is more akin to the gathering of rubbish rather than the assembling of a group of like items. The word "collectibles" has had a connotation within the auction world as goods at the cheaper and less desirable end of the market. Until the last few years, the major auction houses would leave these goods to second tier auction firms. In recent years, however, Art Deco has become a recognized stylistic category and major auction houses regularly offer dedicated sales of high-end Deco pieces.

For those looking for credibility for their pursuits, collecting antiques over 100 years old seems to have an built-in level of legitimacy and status. In the world of art, even modern works are acceptable for collection. But the more modest contemporary objects—particularly if they have been domestic appliances—lack gravitas. They are not only still being remembered in their setting or context, but they are also often rejected because they were simple utilities and worth little at the time. Even the word "collectibles" has a pejorative association, implying the acquisition of objects of lower status. When considering radios in this regard, we collectors are faced with the twin hurdles of appreciating an object that is, first, less than a hundred years old and, second, a cheaply, mass-produced item. Explaining that these radios are an unrecognized and unappreciated embodiment of the Art Deco style has only recently become comprehensible as Machine-age and Streamlined forms that were manufactured in quantity become recognizable today not only as a reflection of an era past, but as part of the design fabric of the industrial evolution that occurred in the twentieth century.

The trend from lowly collectible to valued antique is reflected in the fact that major auction houses are now featuring single owner radio collections in their own right. The Woolley collection (Bonhams, New York, 2007), Salmon Collection (Bonhams, London 2009), the Szental Collection (Joels, Melbourne 2010), the Schoeman Collection (Bonhams, New York 2011), and the Balsbaugh Collection (Bonhams, New York 2012) were all very successful sales and engendered much media publicity and spirited bidding from buyers all over the world. The majority, as well as the most important lots, were radios of colored Catalin and Bakelite. The high prices achieved at these auctions have prompted articles in art and financial publications, which, rather than reporting an anomalous occurrence, are now reinforcing the continuing popularity and value of the quality Catalin and Bakelite radio market.

The true collector is not an investor for profit. Objects are bought for love and the cost is often irrelevant. Granted, it is gratifying when the pieces you bought for very little become valuable, but the emotional relationship between collectors and their possessions tends to impede relinquishing them for money. Many collectors trade up or swap, but this is more about refining the collection than capitalizing on it. Some collectors aim to donate their collections to institutions or give away items to others, but this tendency often decreases in direct proportion to the increase in value of the collection. The idea of gifting a collection to a major institution can be a sad delusion—the art galleries and museums rarely have the inclination, much less the space, to place whole collections on permanent display. Often a few pieces may be seen by the public, while the bulk are stored and often dispersed. The only strategy guaranteeing long-term public exposure would be to provide the institution additional funds for the space and maintenance of the collection, a task beyond almost all but the most wealthy in our society. Most collectors I have met are concerned regarding the final disposition and dispersal of their pieces, as often most around them are not participatory or appreciative, but they often resist parting with the best items since, in some ways, these define their existence.

Some collectors view their possessions through the prism of how and when they were acquired. In distinction to objects bought for their intrinsic historical attributes, this approach is a much more sentimental and personal way of presenting a collection. Of course, the stories of chase and capture add excitement and vividness to the narrative, but the item should still fulfill the inherent values of rarity, excellence, and suitability. The collecting of mementos has a different ethos than collecting antiques. In the interior design world display of so-called "collections" has become a fashionable device in home décor. Often, however, the meticulous positioning of identical balls of twine or a wall of old kitchenalia is just about visual impact and grouping and in some ways a caricature of collecting.

One thing that increasing exposure and value does bring is grudging respect from family and friends. It can, by the same token, also cause bitterness and jealousy from other collectors who may have concentrated on another segment of the same field but whose pieces have not become as desirable. The growing fascination with Bakelite and Catalin radios and the disparity in values between these and wood radios is obviously confusing and disappointing to those traditionalists who love radio in the generic sense, as well for those enthusiasts who restore wooden radios. When seen in the light of the Art Deco movement, in which design, color, and styling are both appealing and timeless, the broader context in which Bakelite and Catalin radios are valued underscores and explains the imbalance. There are, of course, some extraordinary wooden consoles and mantel models that defy the trend, generally because of their outstanding styling and finishes. Looking around the world for the best examples of Deco styling in radios, I have added a number of wooden radios to my collection and in this book I have made sure the best of these are displayed as proudly on the page as their plastic cousins.

Once upon a time, the rumblings of the disaffected were occasional whispers and the odd objectionable letter. However, with easy access to global communication, we have seen some unjustified and unpleasant vitriol spread quickly and widely. Every collecting field can trot out their oddities and eccentrics, but sometimes they are madmen and monsters. In Australia, one particularly offensive dealer/collector consistently plagiarizes images and material from other enthusiasts' websites for his publications. As well, he has mastered the capacity of the Internet to disseminate material with derogatory and racial comments about other collectors and dealers. This does, ironically, exemplify the differences between radio and the Internet. Radio is a gated broadcast vehicle, which controls, to a great extent, the content of information disseminated; the Internet is an unrestricted public-address system. While in principle the open architecture of the Internet seems to be a benefit to the community, it does allow uninhibited dispersion of all manner of data and images, often well below a reasonable level of morality and intellectual rigor.

Collectors should also be very aware that there are a very few unscrupulous dealers and sellers who continue to misrepresent their goods and occasionally fail to send them to the buyer. While *"caveat emptor"* is clearly a maxim to follow, much of the advantage in Internet buying is in the hands of the seller. The advent of market sites (such as eBay) has done a great deal to democratize and make accessible the buying and selling of goods by the public. In the case of radios, it has created a truer market value as well as an international arena. Unfortunately, although eBay (for example) does ban sellers with multiple negative feedback, dishonest operators simply reappear with a new eBay name and continue their activities.

These concerns are outweighed by the thrill of finding new treasures and the excitement of receiving them. One has to accept the occasional disaster as part of the fabric of collecting. The incessant search for something special is often a journey into the unknown—always exciting and generally fulfilling. As well, collecting can be a chance to meet others of similar persuasion, people who love the same things you do, are fascinated by the minutiae that would bore everyone else into catatonia. My previous book, *Radio Days—Australian Bakelite Radios,* involved a series of Tuesday evenings over about six months with three other collectors. We would gather radios together, photograph them on the kitchen table, then argue over the resultant image on the computer screen in terms of orientation and color. All the while we would drink coffee and eat cakes bought from the *patisserie* across the street. This time for all of us remains a pleasant, enduring memory of camaraderie and achievement underscored by our shared love of radios.

An Art Deco polished metal ensemble from the mid-1930s, including a Stewart Warner "Good Champion" radio, a Zephyr clock designed by Kem Weber, a Saturn lamp, and bookends by Zim

Wall of Radios 2014

In most countries, the majority of radio enthusiasts are technologically oriented, with a passion for circuit diagrams and valves. The local and national historical radio organizations generally reflect this preference and, as well, a fair proportion of their members enjoy rehabilitating the wooden cabinets that surround the chassis they have so assiduously reassembled. Collecting is to most a secondary consequence of the mechanical overhaul and rarely an end in itself. Those who simply collect radios for the entity itself often have no great interest in the inner workings and thus tend to have little in common with most radio enthusiasts, remaining somewhat isolated in the radio world.

Of course, one of the great thrills for a radio collector is to find a rare and desirable radio in an unlikely place and at a knockdown price. In Rio de Janeiro in 2011 at the World Art Deco Congress, I was fossicking through the antique marketplace in Copacabana and passed a shop that specialized in old, redundant movie equipment. On the floor below the front window and almost invisible from view were two American Air King radios from 1933 … and one of them was a green export version (four knobs, p. 119). Green Air Kings are in the stratosphere of collectible radios and possibly only 20 or 30 are known. Only the red Air King is more valuable, but we'll leave that one for a magic discovery on another trip.

Unfortunately, both the Rio Air Kings had cracks, but generally the cabinets were complete and capable of being restored. Much haggling took place between the shop-owner, who spoke no English, and myself, the desperate, although outwardly nonchalant, purchaser who spoke no Portuguese. Fortunately, in the end the price that pleased the owner was a fraction of the price I was willing to pay so we were all very satisfied. Within a few minutes I was led to another shop where another gem surfaced: the black and white marbleized Catalin Automatic Tom Thumb on page 222, which was in perfect condition. This time I paid a little more than I needed to, probably because I felt guilty about the price I had paid for the Air Kings. I immediately had the Air Kings packaged and sent to a dealer friend in the USA who set about restoring them both. The green one I would keep and the white I would trade up on another Air King I wanted.

The upshot was that for the remaining three days in Rio it was like Hugh Heffner had come to town looking for bunnies. I was called back to the antique market to see all sorts of radios that had been unearthed now that the big buyer was in town. On one occasion in a restaurant I was asked to go outside and check out a radio on the back seat of a car. It all felt a little unreal and somewhat clandestine, but ultimately just good fun. Needless to say, after the first flush of luck, everything I saw was pedestrian, ugly, and overpriced. I carried the tiny Tom Thumb with me around Brazil and Argentina and showed it to anyone who unwittingly expressed an interest in old radios.

As a collector, value is a bonus, not a core consideration. I see the green Air King as a treasure recognized and a major boost to my collection, a mark of my commitment to the preservation of history and the story of its acquisition a wonderful anecdote proving that collector's fantasies do come true. The Art Deco World Congress is in Shanghai in 2015: my expectations are high. Be ready, citizens of Shanghai, that mad radio collector is coming.

Of course, one thing that earmarks a committed collector is that the collection is never complete, and there is always another wonderful, unique object that cannot be resisted and must be acquired. Actually, most collectors change their focus over time, and one of the great challenges in assembling a representative group of objects is to define the limits and standards that one applies.

The engaging focus of the committed is sadly missing in the remit of those who collect because it is profitable or fashionable. And in the last few years in America and Australia, investors have begun to target Catalins, colored Bakelites, and mirrored radios. This development may put upward pressure on prices and squeeze out true lovers of radio, but once radio assumes the same status as antiques and artwork this is inevitable. One potential benefit of investor-driven collecting is that those radios amassed in bulk in a short period may return to the market within a few years as investment strategies change or financial crises provoke the liquidation of collections.

A book such as *DecoRadio* can obviously serve as a reference for traditional radio collectors, hopefully adding to the pool of knowledge about the subject. This is a celebration on one hand and on the other a formalizing of the theme and field of interest. *DecoRadio* is presenting radio as a global phenomenon, but of course most collections will predominantly contain local or national examples and reflect a natural desire to highlight one's own country's achievements. Collecting locally is also easier in terms of availability and cost. With the advent of Ebay and specialist sites on the Internet this is slowly changing as access to international radios becomes easier and a growing number of collectors are adding radios from countries other than their own.

It is also clear today that these radios are of interest to a much broader group of people than the original radio enthusiasts. We are seeing individuals and organizations collecting radios because of the industrial designer connection or Art Deco styling, and I believe this will become more pervasive and eventually change the status of radio markedly and permanently. Some people collect because they like the retro and trendy look of these old radios. Who is to say that is not a worthy motivation?

Of course, not everyone wants to be a collector, and many more people can appreciate the style and variety when a group of objects are brought together. This is where museums, galleries, and exhibitions act as interface with the public and provide the setting to properly display multiple works of art for the community to enjoy. Sadly, radios have not been given much prominence in these venues, but at some stage an organization will bravely showcase the best of radio. It is worth mentioning that in the UK and Australia exhibitions of Art Deco objects have always engendered enormous interest and record numbers of visitors. Wherever radios have been included, these have proved especially popular.

In the absence of (or associated with) a physical event, books can bring collections to people. Granted, it is not the real thing, but at least it is a permanent record, whereas exhibitions have a limited time-frame for viewing. It is another reason why a book such as *DecoRadio* should aim for the highest standards of photographic reproduction—often the visual material is absorbed and appreciated much more quickly and profoundly than the text. The text may contain the substance of the argument, but the images are immediately compelling and often sustaining.

It is often difficult for a collector to articulate the rationale or grand purpose that drives the acquisitional fervor. For some it is just the thrill of the chase, an addictive adrenalin rush about winning the auction or ferreting out a bargain in a junk shop. For many of us it is an emotional and impulsive quest to have around us objects that feed our soul, which we love to see and touch and that seem more meaningful in the presence of their brethren.

I like the Japanese idea of "wabi sabi," which is an appreciation of the imperfect and the impermanent. It releases us from revering only that which is perfect and infinite. Items that were commonplace and utilitarian in their time can be better appreciated through the twin lenses of time and wisdom with the best gathering respect rather than dust. No matter the theme, the foundations of serious collecting must always be based on quality, significance, suitability, and passion.

I see these radios, and their numerous design iterations and countless color variations, as iconic representations of the Deco period and style. They inform us of an important nexus of social and industrial change in the 1930s and 1940s. But that is an intellectual explanation and on a much more visceral level I just love their shape, color, and feel.

CHAPTER 14

REMEMBERING RADIO

Almost 100 years since its inception, radio is still a force in entertainment and information, but since the 1950s has ceded influence to television and now is beginning to become part of the digital world of the Internet. In a few years, one can imagine that radio will not exist as a separate box attached to an electrical outlet but will be only accessed with a computer and Internet connection. With improvements in Wi-Fi, radio will finally be truly wireless.

One distinct advantage of modern digital radio is the ability to retrieve archived as well as current programs, whereas older radio listening was always an immediate and momentary opportunity. Part of the excitement of yesteryear was waiting for a program at a particular time-slot and not wanting to miss a one-time occasion. Radio has always been about entertainment, drama, sport, and information, with each country developing its own particular programs and personalities. Radio still functions in this way, but whereas once it was the only source of home entertainment, today it has competition from television, DVD/video, and the Internet.

Using computers and the Internet to download music and radio programs has changed the consumer from the passive participant of yesteryear to one with exceptional capacity and flexibility to choose what he or she listens to and when. On the other hand, attention span appear to be shortening with each succeeding generation and the sound byte is contracting in tandem. People listen to a single track of music rather than an album, and rarely want extended, authoritative views on a subject. In the post-modern world, just an opinion (and it matters little whose) seems to suffice.

For baby boomers and those born earlier, the valve radio is an established memory. We had one in the home or at our grandparents' home. It is easily recognizable in all its iterations, in sizes ranging from console to tabletop to portable. It was a box of some sort with a dial displaying radio stations, a speaker, and at least two knobs. While there are still new radios offered in a retro style, things have changed from analogue to digital. The dial is an LCD display allowing finger-tip access to all major operations, the knobs are buttons which have preset and variable functions, and an FM tuner has been added.

Radio and television were in many respects competing systems, one aural and the other visual. Television retains an important feature that identifies it as a visual medium—the screen. In fact, as technology has shrunk the television set, all that is left today is the screen. TV may now be part of a computer system and accessed via the Internet but it will continue to be recognizable. Not so for the radio—as an audio instrument, its speaker was initially an independent element in the 1920s, but since then has been subtly incorporated into the fabric of the radio and is rarely a dominant feature. People have become less and less concerned with the quality of sound reproduction. Digital recording has compressed sound data and consequently speakers do not have to be of high quality. Thus, radio does not now require a dedicated vehicle and speakers can be small and inconspicuous. In the future, radio will more and more be an Internet option and less an object in its own right.

Children born today will not think of radio as furniture or as a domestic appliance. That said, radio continues on regardless of how it is transmitted and received. There will always be people wanting a form of entertainment that can be absorbed while other work is being performed. One can listen to a radio while driving a car, ironing the clothes, digging in the garden, and working on the factory floor. My children claimed they were able to study while listening to the radio although I doubt if the academic uptake was maximal. None of this works with TV, so perhaps radio still has a long-term future.

Am I being old-fashioned and whimsical to think that these radios of yesterday should be remembered? I believe that respecting history and preserving our heritage is a mark of humanity and a responsibility of every member of the community. This applies whether it is an old building or an old radio. Truly, many of both are unremarkable and should be recycled or adaptively re-used, but the special ones inform us of the past—how things were made and how they were used. These should be treasured, restored, and displayed for all to enjoy. Sadly, most people are never going to see these radios in the flesh and a book is one way to widely disseminate this information, albeit a poor alternative to reality.

For those who recognize and appreciate radio, the gems in this book will be a reference and a souvenir. But for our children and grandchildren, *DecoRadio* speaks of a time when radio was one of the most important sources of information and entertainment. It will serve to give them a sense of the history of what was then a discrete form of communication, and today is part of a blended barrage of modern broadcasting. Radio still serves as an arbiter of social exchange, but as it dissolves into the cocktail of modern mass media, it is surely important to articulate the evolution of its social, technical, and design manifestations. The best of these radios will be randomly preserved in the radio diaspora, but here, for one moment, they are all united.

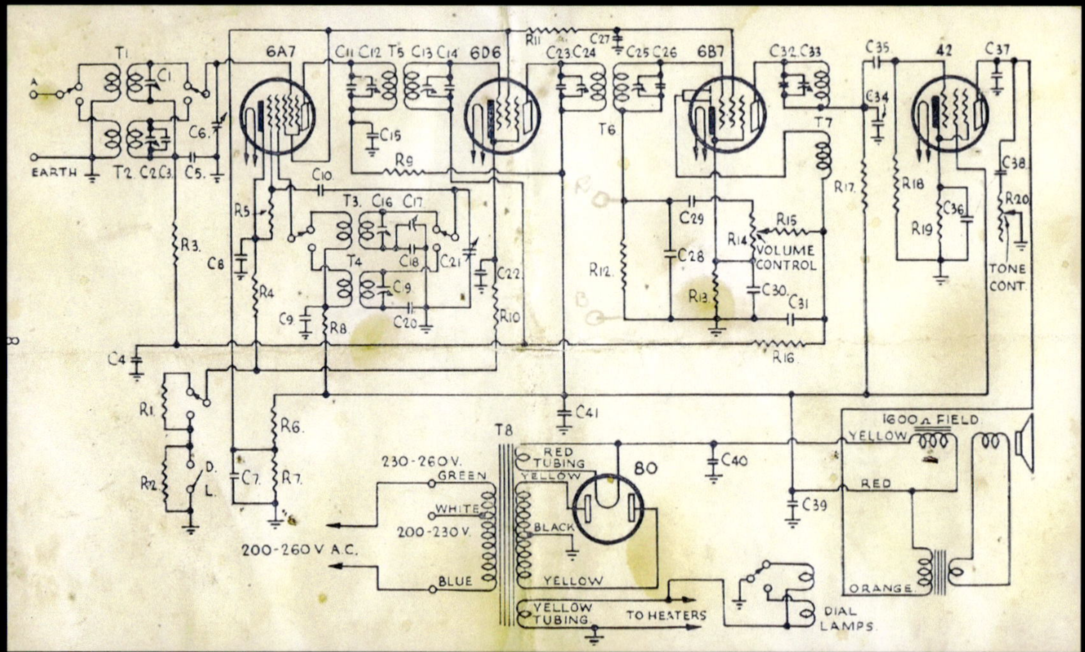

CHAPTER 15
CONCLUSION

The radio can be viewed in the context of its design aesthetic without discounting its validity as a medium for communication. In this sense, radio becomes the hub of a wheel with spokes to Art Deco, Streamlining, industrial design, Machine-age processes, and modern materials. In utilizing an array of novel and traditional shapes and materials for the radio cabinet, design professionals interpreted a modern stylistic movement through the vehicle of a mass-produced domestic appliance and with global ramifications. The time frame for the development of the valve radio parallels that of the Art Deco movement, both beginning in the mid-1920s and fading in the early 1950s. Modern styling and industrial design evolved and matured in Europe, the USA, and UK in the 1930s and radio styling all round the world reflected these developments.

To validate a thesis one cannot simply point to a couple of examples and expect this to be sufficient. But here there is sufficient power in the samples presented to validate and support the core hypotheses. Thirty-five designers have been identified (including most of the principal instigators of Streamlining) who created over 90 different modern styled radios. As well, *DecoRadio* presents a broad selection of over 300 outstanding radios from 15 countries, all of them being examples of a worldwide phenomenon in design and reflecting a whole array of local interpretations of the Deco style. The story and background is important, but the radios are visually such great ambassadors that they actually do not require much championing.

The vast majority are mantel (or tabletop) radios, because, from 1930, this is the novel form that allowed designers full expression of modern Streamlined styling with mass production, using new materials and an array of colors. However, I have included eight console radios from Australia, France, New Zealand, Belgium, and America, all with particularly fine Art Deco attributes.

I started with a view that all wood radios were mainly old-fashioned and for many years lagged behind the more modern plastic cabinets. However, I found that a number of companies produced beautiful wood console and tabletop radios in the early and mid-1930s. A few are shaped with elegance and confidently adopt modern styling, but most incorporated fine color variations in the wood veneers as well as decorative motifs in wood and metal, which gave them substance and individuality. There may be only eight consoles and nineteen wood mantels in the book (and some of the latter are almost as big and as heavy as the floor-based consoles), but they are masterpieces in the Art Deco style and make a material contribution to the book and to the story.

While radio was fundamentally a vehicle for communication, it also had a role during the 1930s and 1940s as a piece of furniture, a domestic appliance, and part of a new style of modern object in home décor. The mantel radio in all its color and shape iterations extended the role of radio to different members of the family and in multiple sites in the home and work setting. Radio, unlike the car and the telephone at the time, which tended to be one per family or household, became available to the individual and particularly women and children were emancipated and empowered by their access to programs that appealed to them and could be listened to with their own personal radio. The Streamlined radio cabinet represented, and was symbolic of, the new machine-age future and this was by virtue of its aesthetics, independent of the core audio function.

A beautiful radio can satisfy all the senses. Looking at the colors and shapes is just a visual delight. Running your hand over the surface of a Catalin or Bakelite radio, the smoothness and fluid curves are almost sensual. When you turn on an old valve radio, there is first nothing, then a hum, and then the crackle of static. These aural cues speak of a time past and of another social milieu.

It may be a cliché, but it is important that we acknowledge history and then conserve it. Here I hope I have shown the influence of industrial designers on the Art Deco movement via the medium of radio. Though this association has not had much exposure or traction in the radio, design, or Art Deco literature, it is fortunate that collectors of radio have preserved the best examples, generally with little knowledge of these influences. Hopefully, now with a recognized provenance and heritage, these radios will be better appreciated by the public, by collectors, and by historians. Time is a two-edged sword. Objects over time deteriorate and are cast-off while the increasing years create antiquities. The 1930s and 1940s are becoming history rather than memory and this will confer increasing importance on some of the objects of that time. It is easy to venerate the expensive elitist pieces, but much more challenging to recognize and preserve the common objects like radio, which, nevertheless, changed the world.

RADIO GALLERY

1920s

Germany

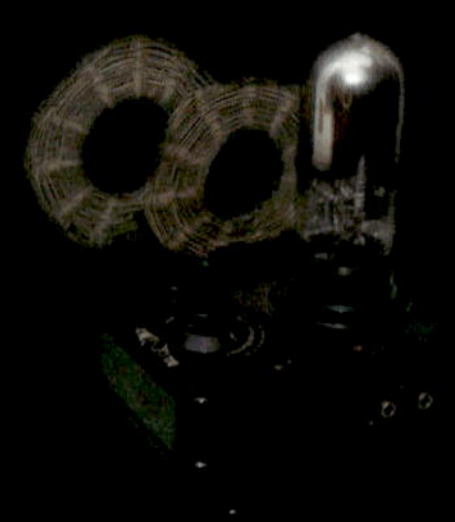

France

UK

Netherlands

REXOPHONE, UNKNOWN

Crystal Sets

UK

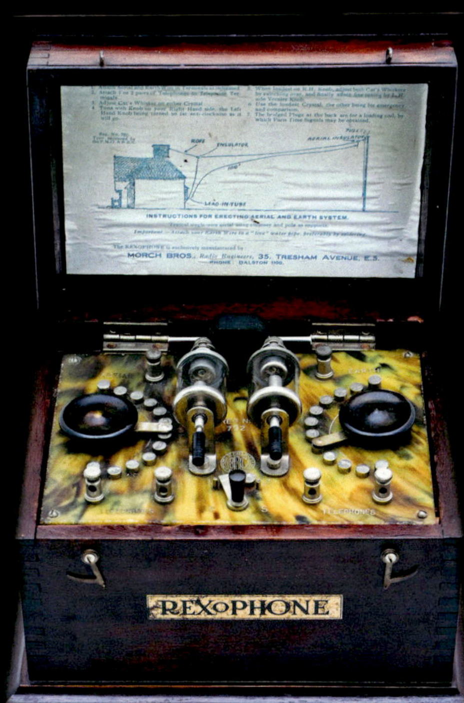

1926 LOEWE OPTA OE433
Germany

"pas de foyer
sans poste
PHILIPS"
une prise de
courant et...
c'est tout
AFRIQUE DU NORD
PHILIPS

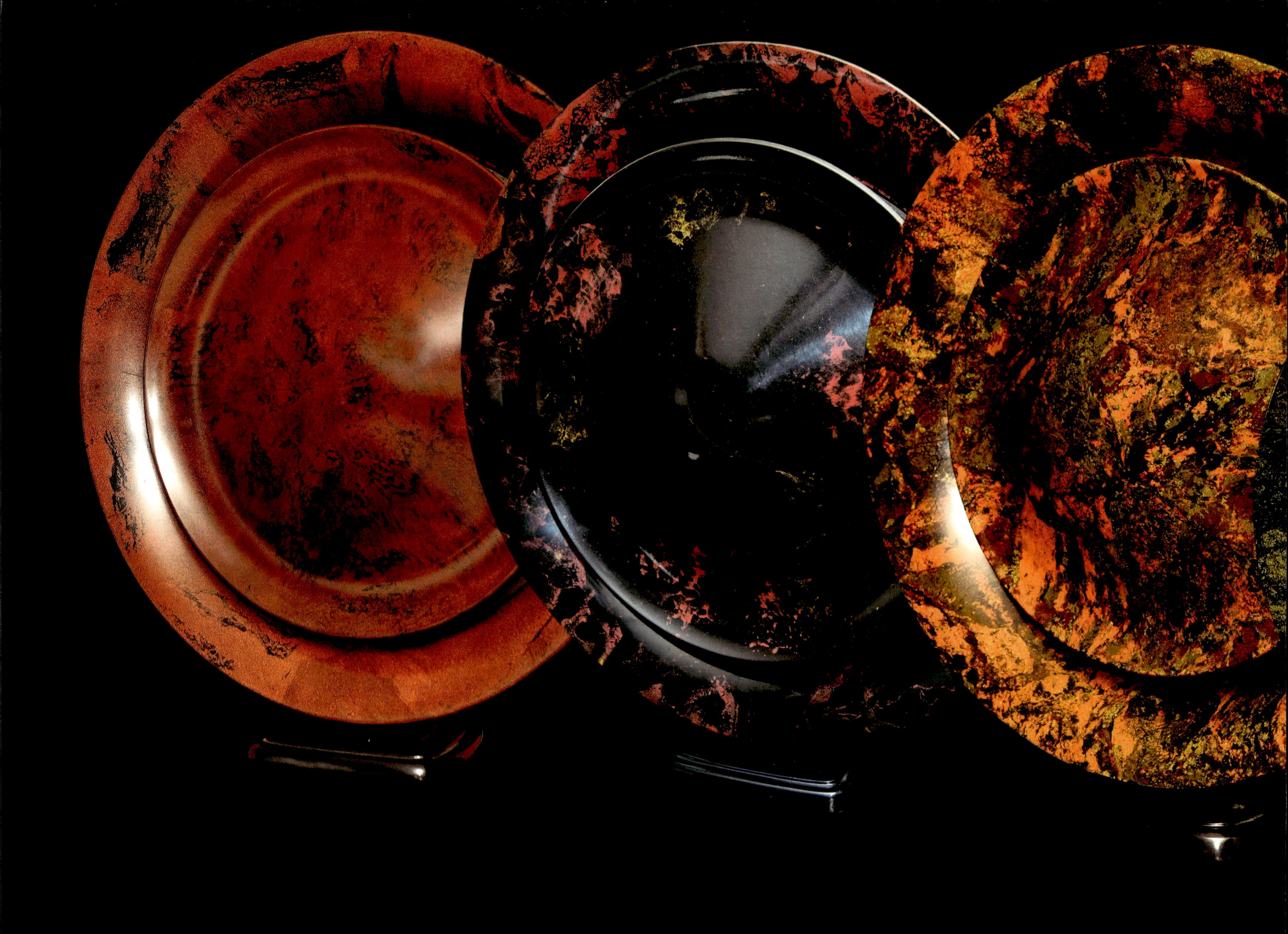

PHILIPS PCJJ Loudspeaker 2007/2003/2015 **1927**
Netherlands
Designer: L C Kalff

1927 HERVOR "Minilux" Loudspeaker
France

LA SCIENCE ET LA VIE
La nouvelle formule de diffuseur
FAIRE PETIT
mais
SENSIBLE
NET
PUISSANT
MINILUX-HERVOR
n'a que 8 cm sur 13 cm et
se fait entendre
HERBELOT et VORMS
Constructeurs
35. RUE DE BAGNOLET - PARIS (20e)
Téléphone : Roquette 50-13 et 22-59
publix
Son prix le met à la portée de tous. En jolie boîte écrin. 6 modèles au choix.
200 frs

1929 PHILIPS 2113 "Permagnetic" Loudspeaker

Netherlands

Designer: Louis C. Kalff

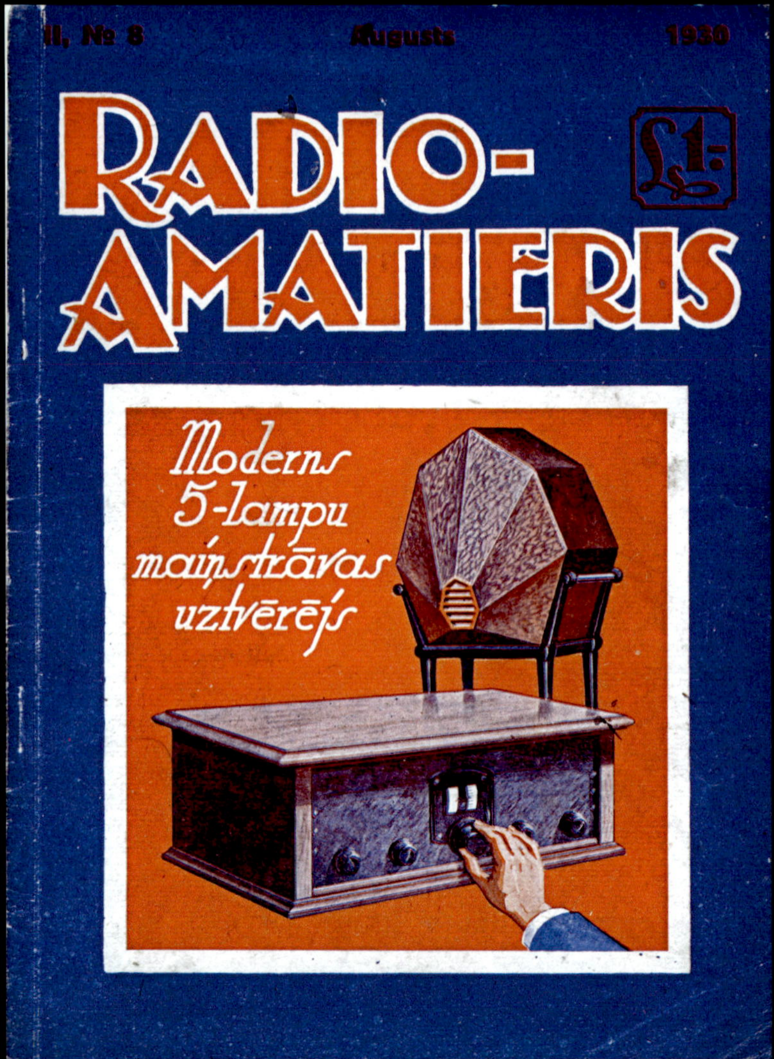

NORA K3W
Germany

1929

1930s

USA

USA

1930s

USA

1930s

1930s

USA

USA

GERMANY

UK

1930s

USA

GERMANY

UK

UK

1930s

BELGIUM

NETHERLANDS

1930s

BELGIUM

ITALY

AUSTRALIA

1930s

AUSTRALIA

AUSTRALIA

FRANCE

1930s

AUSTRALIA

FRANCE

NEW ZEALAND

1930 NORA W3L "Sonnenblume"
Germany

KOLSTER BRANDES "Masterpiece" 1930
UK

• THE GIFT OF THE YEAR •

THE INTERNATIONAL KADETTE Radio

Here's a marvelous little radio so compact it easily hides away in a desk drawer —yet is ever ready to pour forth programs of information or entertainment.

Moreover, it is the only radio that operates on any 110-volt circuit—either AC or DC—25 or 60 cycle current.

No matter where you are or where you go, the little Kadette performs perfectly—in your office, home, hotel or aboard ship.

Weighs only 5 pounds—so light it carries like a camera or finds a ready place in your traveling bag.

Employs full size tubes—giving amazing volume—crystal-like clarity—and exceptional tone quality. Has self-contained aerial and requires no connecting other than plugging it into the light socket.

Cased in genuine bakelite, black, mahogany or walnut—harmonizing with office or home furnishings of the highest character.

The Kadette Radio is sold by better stores everywhere. Write for descriptive circular and names of dealers in your community.

Price **$25.00** *complete with tubes*
Carrying case $1.50 extra

• *A Gift Idea*—The Kadette Radio offers a novel, practical and pleasing answer to business men seeking a gift for important clients, branch managers, zone managers or salesmen. Correspondence invited.

THE INTERNATIONAL RADIO CORPORATION
27 William Street, Ann Arbor, Michigan

KADETTE H "International" **1931**
(Gothic)
USA

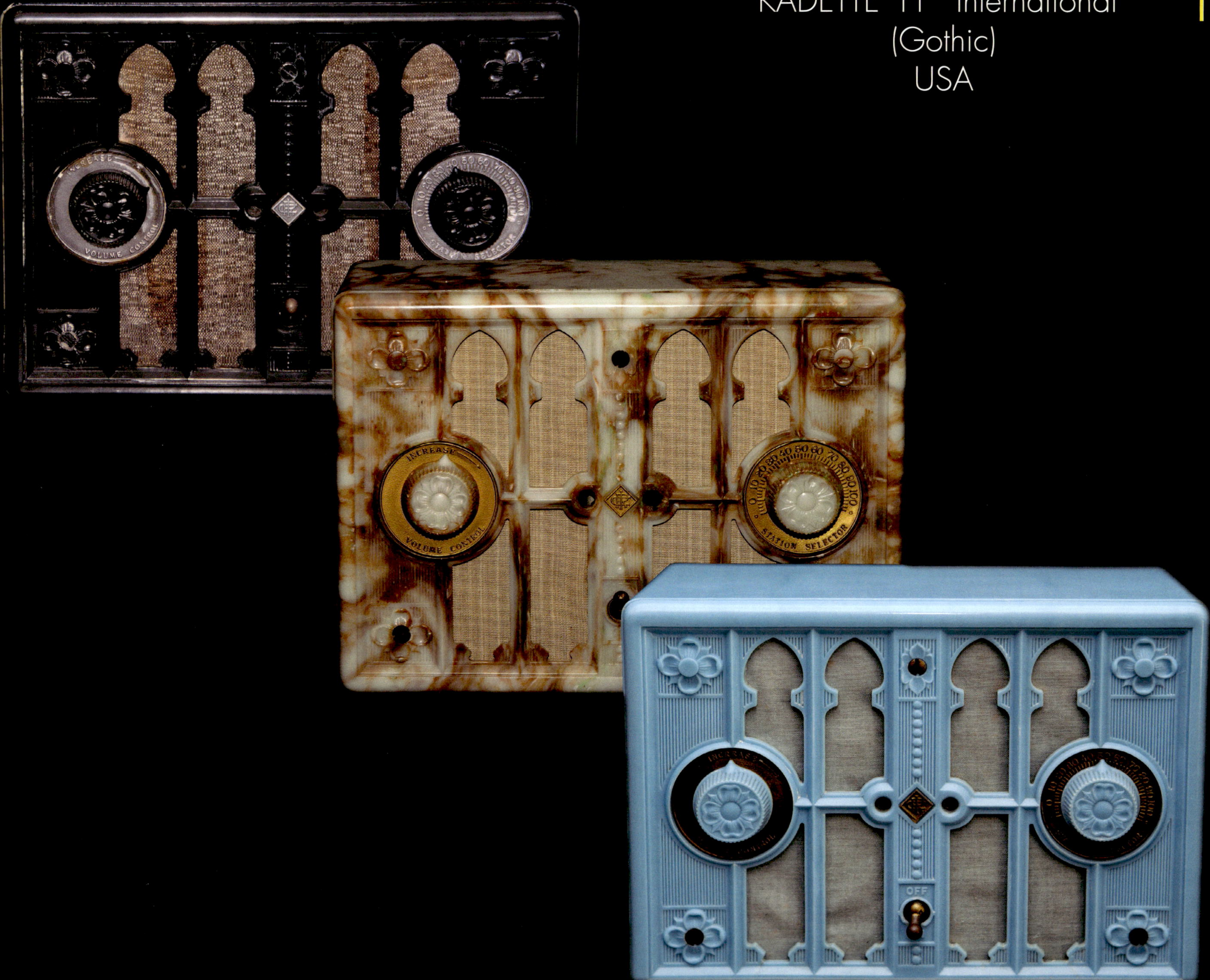

The FISK RADIOLETTE

A SMALL RECEIVER OF BIG PERFORMANCE

This small and compact but powerful radio set gives exceptional performance, and is efficient in the reception of Interstate stations.

You have but to listen to its pure, clear, life-like tone—to inspect its sturdy construction—to realise how faithfully A.W.A. has carried out its task in designing such an efficient radio in such small compass.

The Radiolette is moulded in one piece by a special process, giving it a beautiful finish.

The Radiolette is attractive in design and colour. It takes no more space on a mantelpiece than an 8-day clock, and is just the thing for the den, the verandah, the week-end retreat—for that second set you have always wanted, it is ideal.

The Radiolette is portable—it can be removed from room to room without any difficulty, giving you a radio set in every room at the price of one. Size 16in. high, $12\frac{1}{2}$in. wide, 7in. deep.

£22/10/-

Easy Terms Available.

Tune in to the RADIOLA HOUR

At 8 p.m. every Thursday
Stations 2SM, 2GB Sydney, 2AY Albury, 3DB Melbourne, 3BO Bendigo, 5DN Adelaide, and 4BC Brisbane.

At 8 p.m. every Tuesday
Station 2UW Sydney.

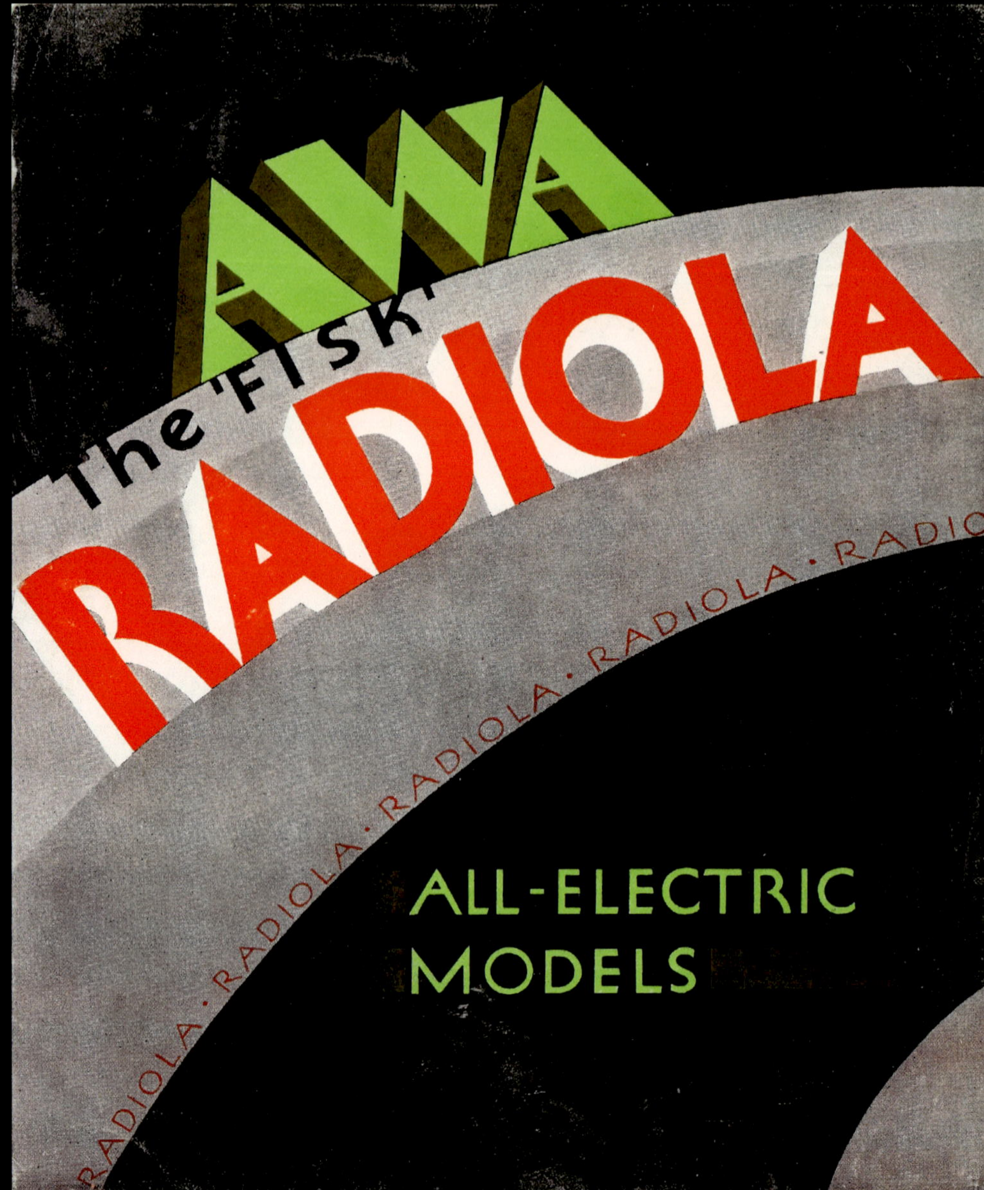

AWA C87 & C104 Fisk Radiolette
Australia

1932/1933

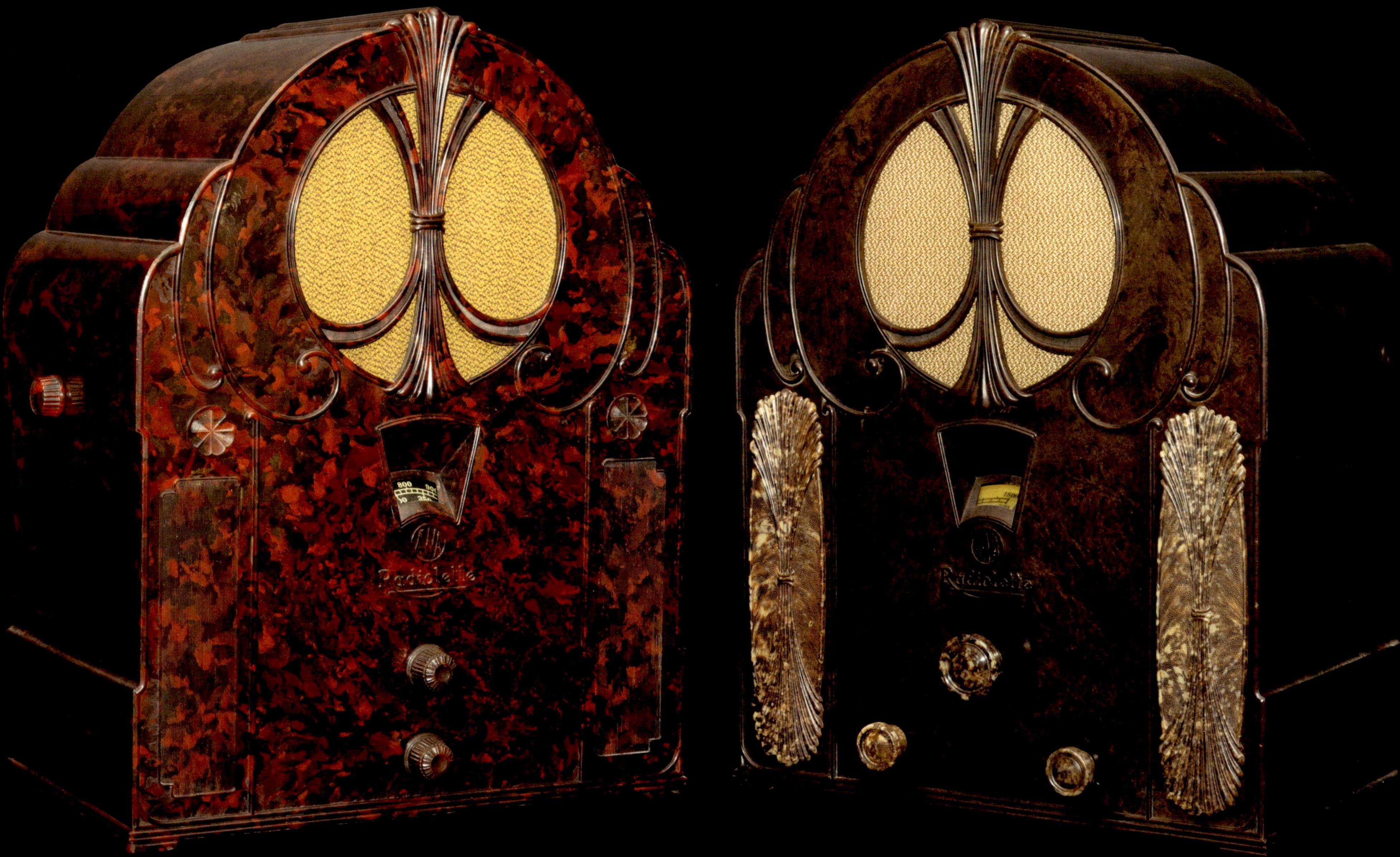

1932 KADETTE F "Junior"
USA

1931/1932

PHILIPS 930C
Netherlands

PHILIPS 830A
Netherlands

1932 EKCO RS3
UK
Designer: J. K. White

PHILIPS 820A
Netherlands

1932

New Air King Set

A modernistic cabinet with a built-in modernistic electric clock make the Air King radio of the Air King Products Co. Inc., 27 Hooper St., Brooklyn, N. Y., a particularly attractive model. It is available in 13 colors including black, ivory, green, red, mother of pearl, walnut, mahogany, etc.

This is a 6 tube superhet covering 550-1650 and 5800-17,500 kc. The tubes used are 6A7, 6D6, 75, 43, 25Z5 and 340. $49.50, complete.—*Radio Retailing*, July, 1935.

Air King Clock Radio

A most striking midget radio designed along ultra-modern lines can be obtained from the Air King Products Company, Inc., 27 Hooper St., Brooklyn, N. Y.

The cabinet is a moulded job, made out of a urea material and may be had in every color of the rainbow as well as black with contrasting colors in the clock and bezel.

The set is a 5-tube superheterodyne using a combined detector oscillator and one intermediate frequency stage, with a two-gang condenser using 456 intermediate frequency. Tubes used are 2-57, 1-58, 1-247 and 280. Speaker is a dynamic and the electric clock on the panel is of a synchronous type. Provision for starting and setting are on the back. This set measures 9½ in. wide by 12 in. high, by 7½ in. deep, and lists at $34.50.—*Radio Retailing*, March, 1933.

AIR KING 52-54; 37-39
USA
Designer: Harold Van Doren

1933

1933–1935

AIR KING 66
USA
Designer: Harold Van Doren

AIR KING
AIR KING
AIR KING
AIR KING

The NEW WORLD DUAL WAVE RADIO

"Old Ivory" Model 702 LUXURIOUS HIGH-LUSTRE IVORY AND GOLD FINISH. ALL OUTSIDE METAL TRIM 22-KARAT GOLD PLATED. SAME TECHNICALLY AS "MEPHISTO" MODEL. R C A RADIOTRONS INCLUDED.

The Ultimate in Design ...

THREE SUPERB COLOR COMBINATIONS

Black and Gold, and Maroon and Gold at $59.50—Old Ivory and Gold at $62.50. Complete with Radiotrons. Prices slightly higher on West Coast.

5-tube AC-DC Dual-Wave Superheterodyne receiver with 7-tube performance in power and natural tone. 3-gang condenser, automatic volume control, double-grille dynamic speaker base—all of unusually substantial, shock-proof construction. Height, 16 inches. Globe and base design covered by U. S. Patent. Licensed under patents of Radio Corporation of America. Fully guaranteed by the Colonial Radio Corporation.

COLONIAL RADIO CORPORATION, BUFFALO, N. Y.

COLONIAL "NEW WORLD" (Globe): "Old Ivory" 702 & "Nubian" 701
USA
Designer: Raymond Loewy

1933

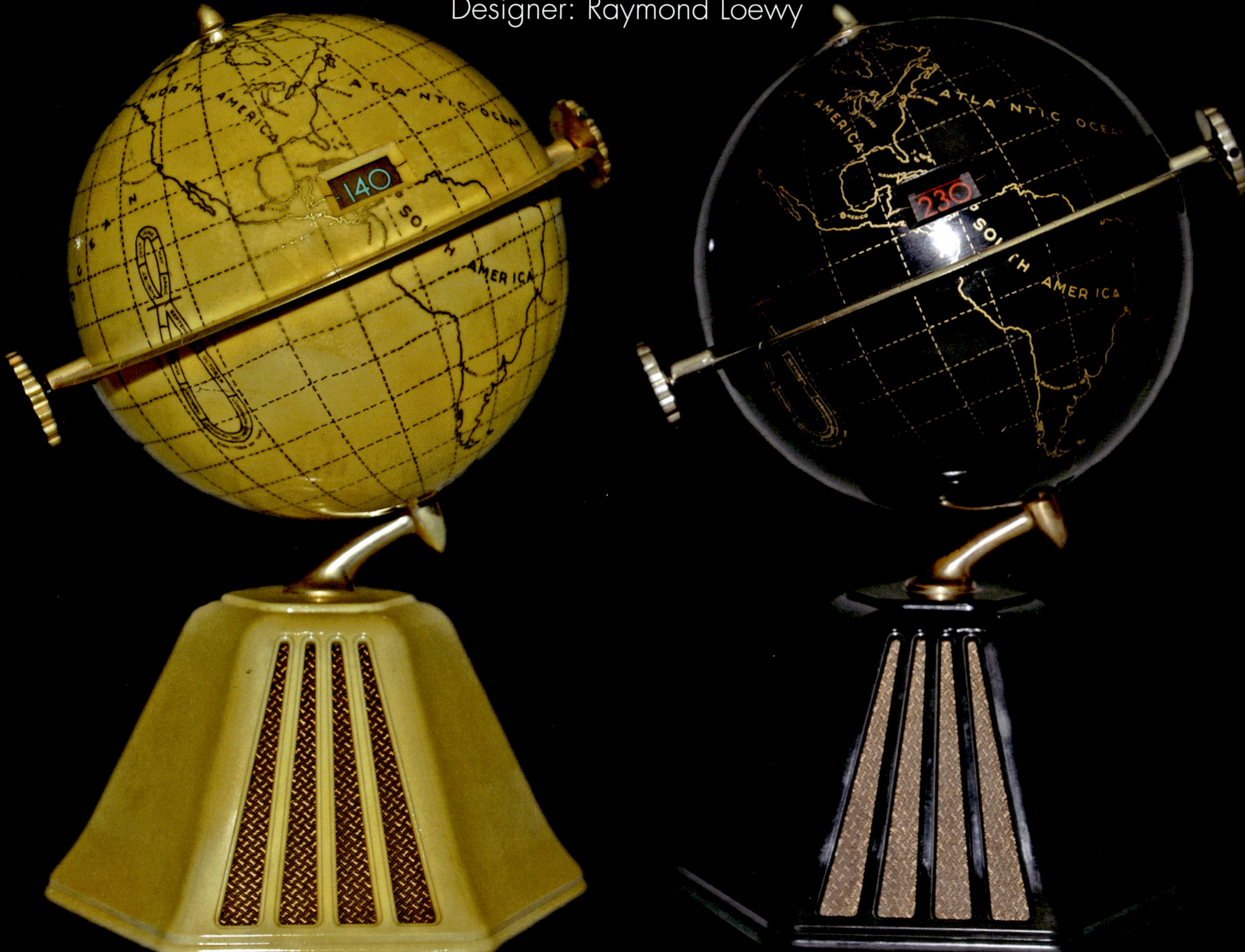

Aug. 29, 1933.
R. G. F. LOEWY
Des. 90,586
CABINET FOR RADIORECEIVER
Filed June 24, 1933
600
INVENTOR
RAYMOND G. F. LOEWY
BY
ATTORNEY

COLONIAL 300 1933

USA

Designer: Raymond Loewy

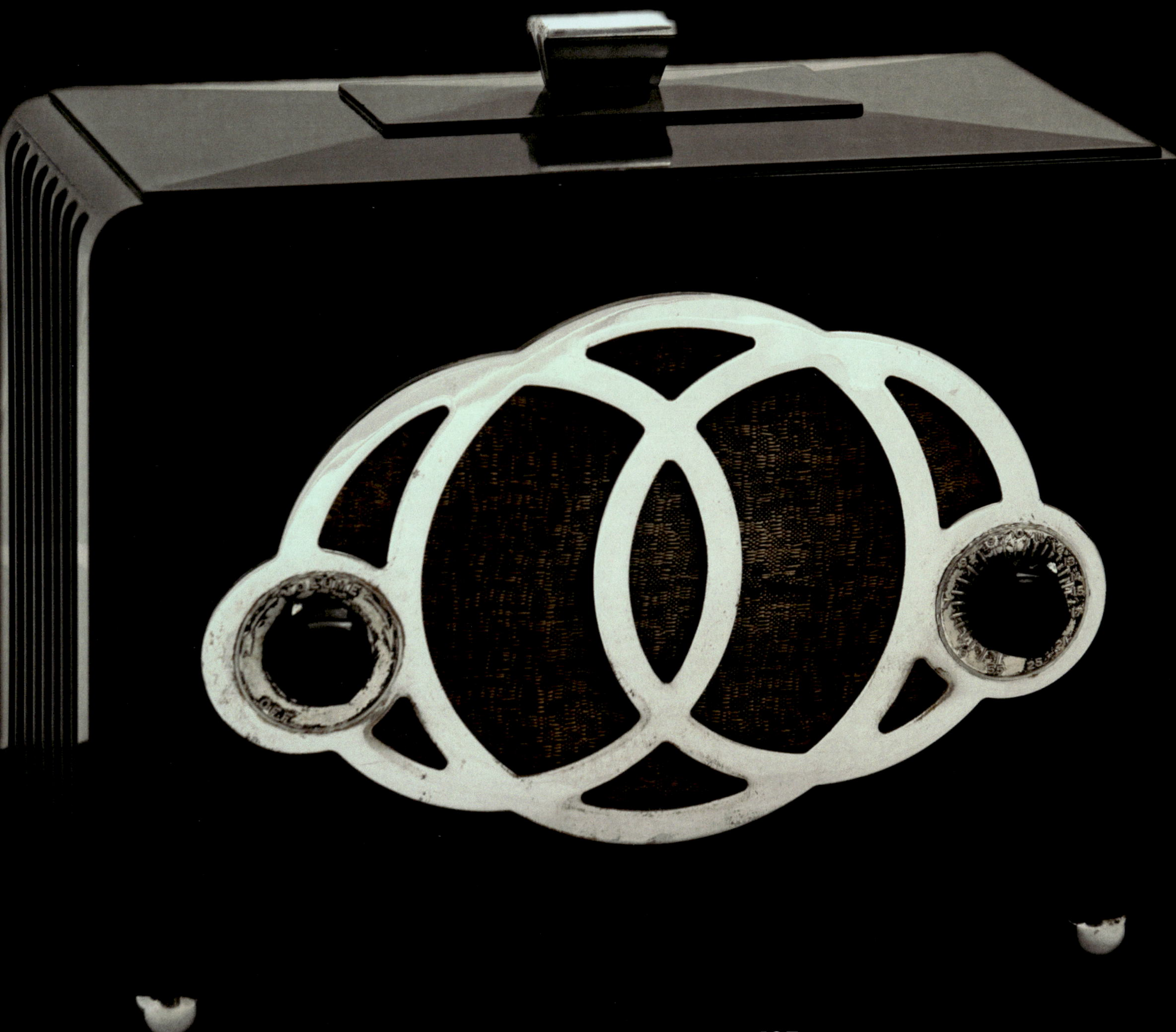

Built for PERFORMANCE

COLONIAL

Really DIFFERENT in Design!

COLONIAL Model 300 $30 COMPLETE

with R.C.A. Tubes and Built-in Aerial

Designed in the popular contemporary style, this cabinet will immediately take the fancy of owners of modern furnishings, or of those who want to add a modern touch to a room. Constructed of satin finished ebony black Durez with polished chromium grille and trimmings. Here is a headliner for window display—advertising feature—SALES—and PROFIT

(Prices slightly higher West of the Rockies)

NOW . . . through the engineering genius of Dr. Fulton Cutting and his associates, every limitation commonly found in compact receivers has been overcome. Here is PERFORMANCE that will satisfy, even under the difficult reception conditions of the summer months. *Consult the nearest of the 76 Graybar branch houses for full details.*

COLONIAL RADIO CORPORATION

Buffalo, New York

1934

1933

TELEFUNKEN 500 (Jelly Mold)
Germany

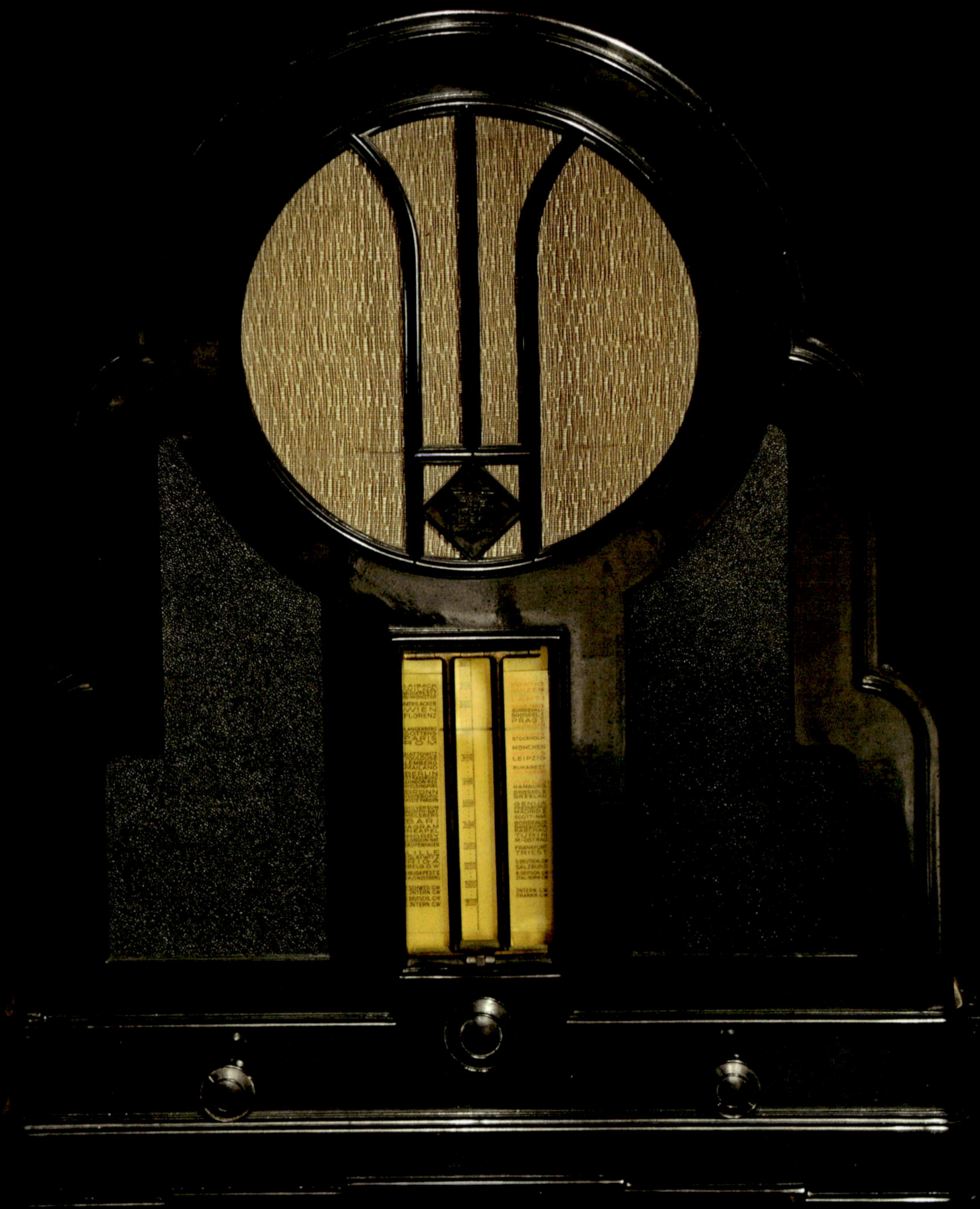

MENDE 148G
Germany

NORA VE301 "Volksempfänger" 1933
Germany
Designer: Walter Maria Kersting

1933 ASTOR "Mickey Mouse" and "Minnie Mouse" Auditorium Speaker Australia

STEWART WARNER 108X (Lady's Head)
USA

1933

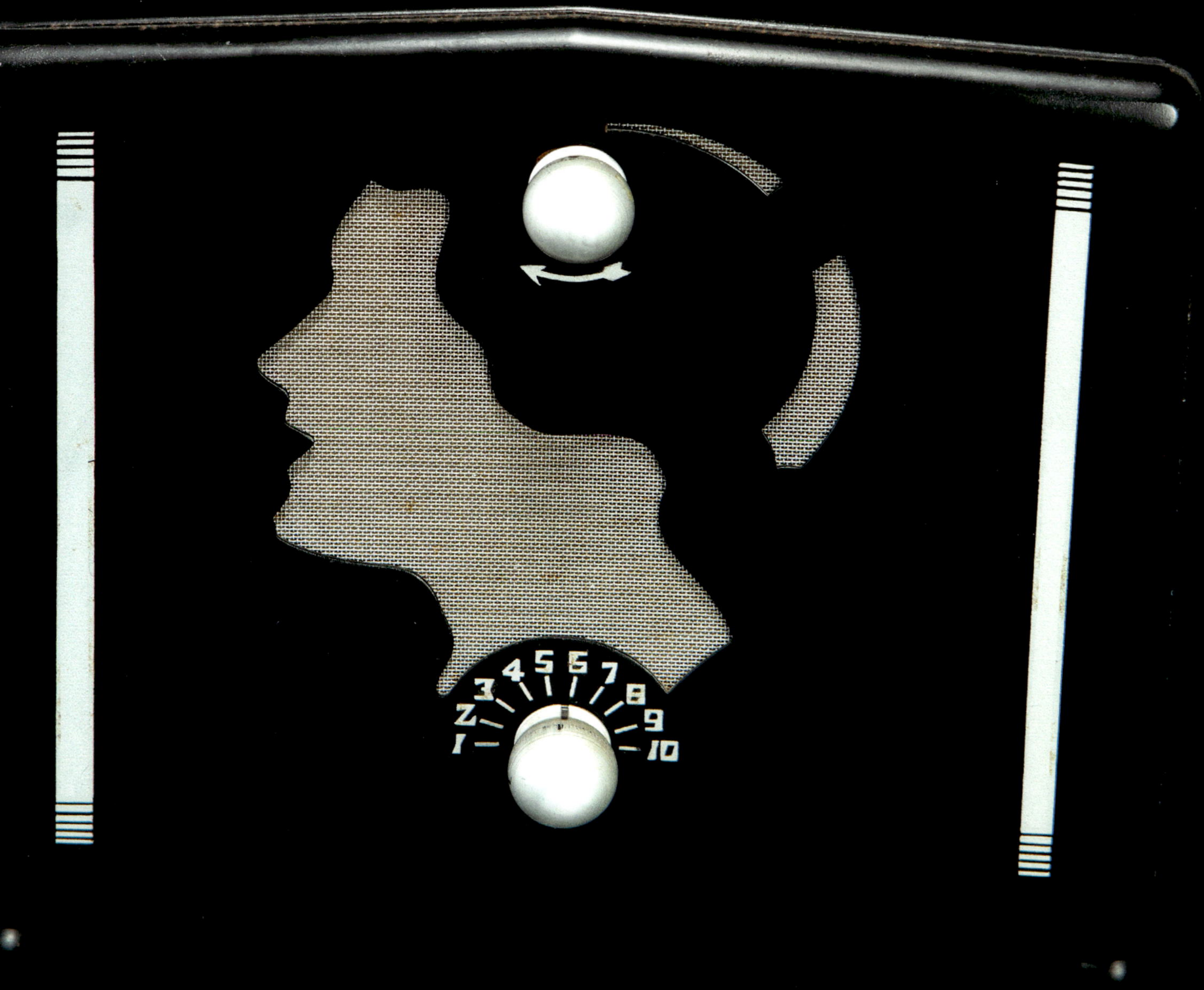

1933

DUCRETET C25
France

ZENITH 1000Z
"Stratosphere"
USA

1933

RUBIS 70 "Art Deco" 1933
Belgium

MAJESTIC GG 463 "Century Six"
USA

EKCO AC 64 1933
UK
Designer: Serge Chemayeff

1933 RCA 115
USA

SONORA S5
France

SONORA Series 3 "Sonorette"
France

1933

MAJESTIC GG 49A
USA

LUMOPHON WD210 "Markgraf" 1934
Germany

1934

PACIFIC "Elite"
New Zealand

EKCO AD65 1934
UK
Designer: Wells Coates

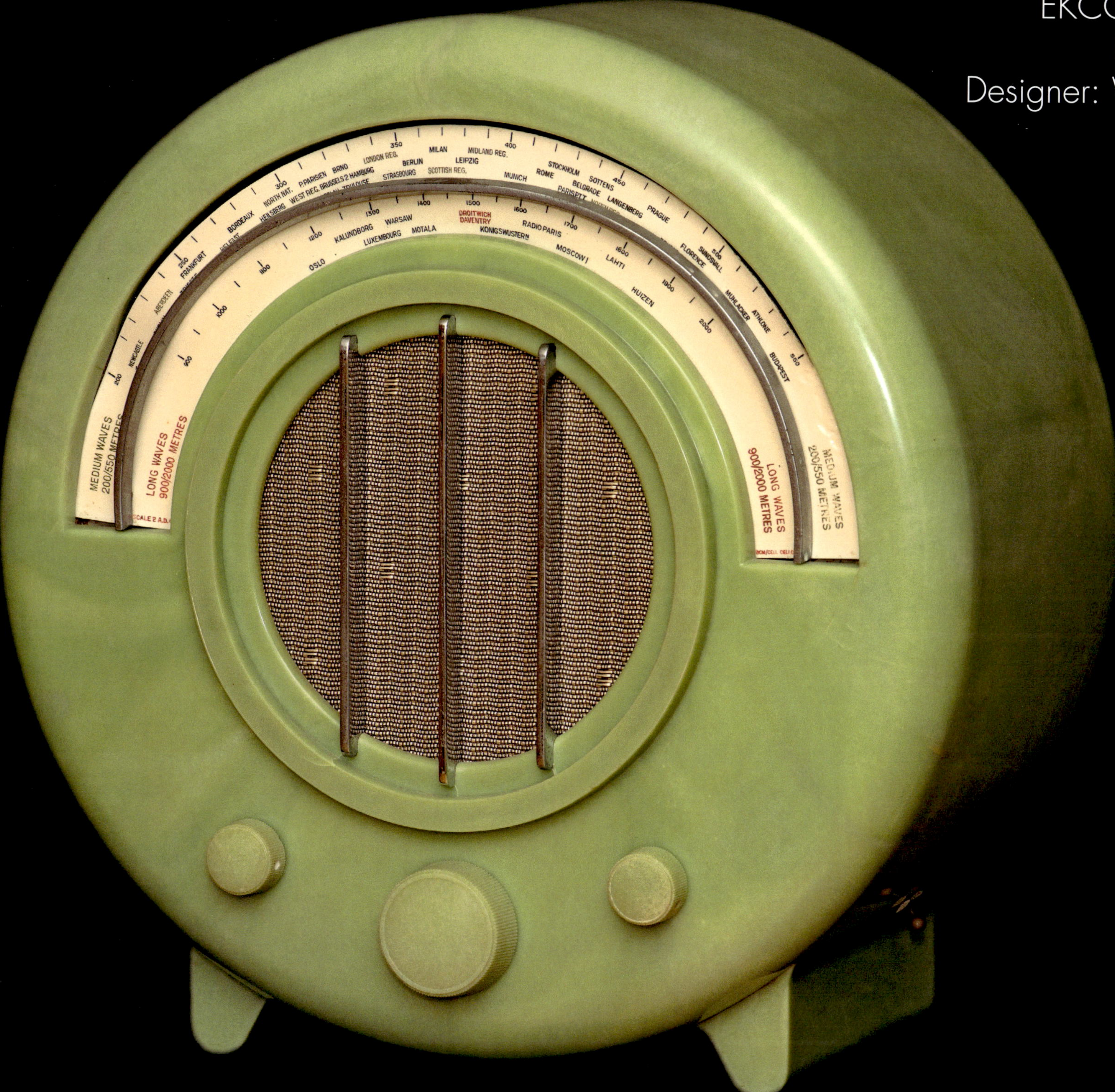

MEDIUM WAVES
200/550 METRES
LONG WAVES
900/2000 METRES

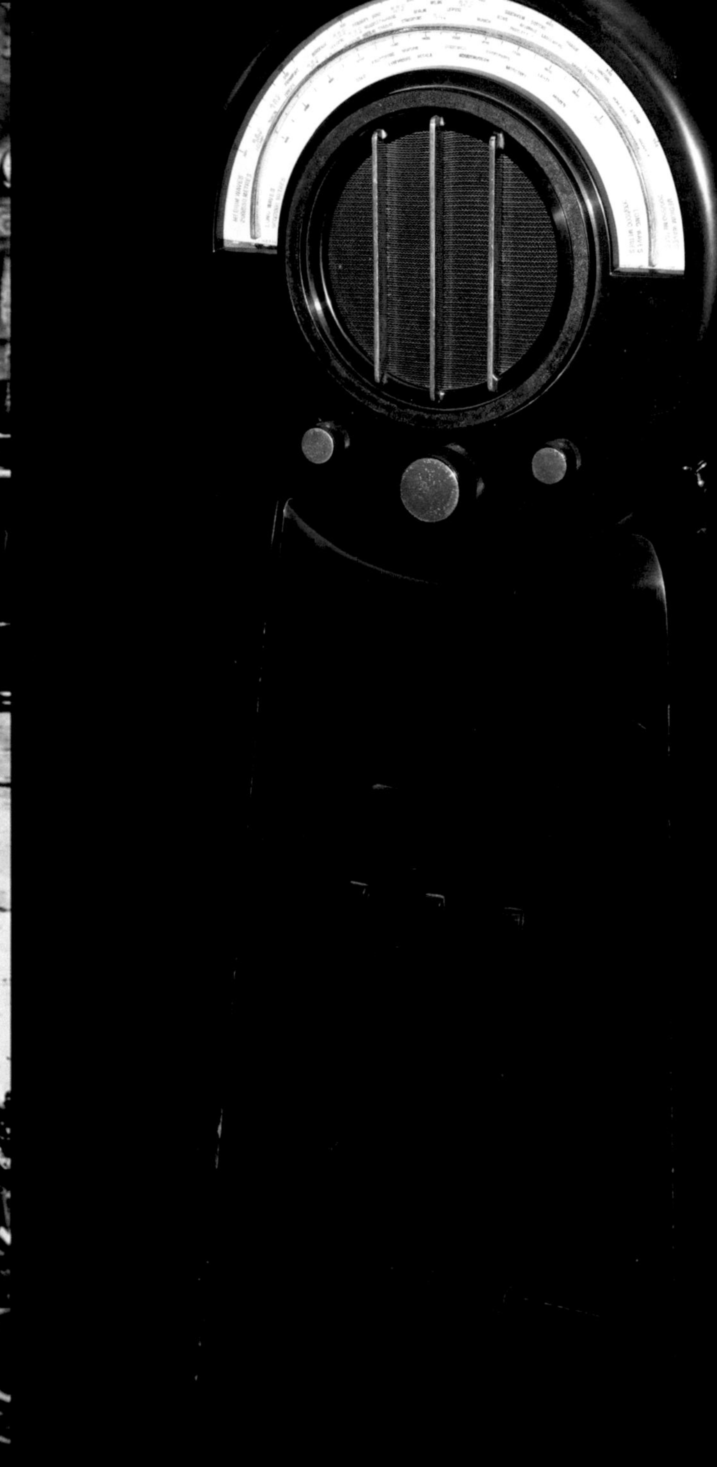

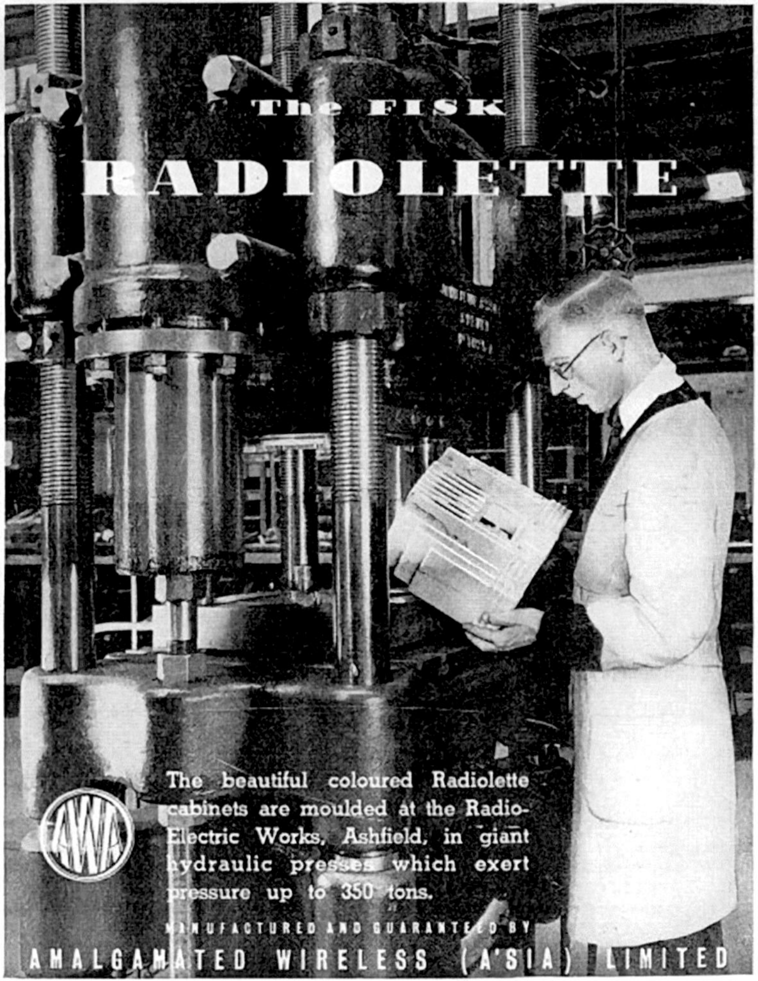
The FISK
RADIOLETTE
The beautiful coloured Radiolette cabinets are moulded at the Radio-Electric Works, Ashfield, in giant hydraulic presses which exert pressure up to 350 tons.
AWA
MANUFACTURED AND GUARANTEED BY
AMALGAMATED WIRELESS (A'SIA) LIMITED

THE SECOND SET IN THE HOME
FOR THE KITCHEN
FOR THE SUNPORCH
AWA
With a Fisk Radiolette you can enjoy the facility of radio in every room. It is easily carried and is the ideal receiver for use in the bedroom, kitchen, dining-room, on the verandah, or at the week-end cottage. . . . A sparkling array of beautifully coloured cabinets, which harmonise perfectly with modern interiors, are available. Prices ranging from 15 guineas.
The FISK
RADIOLETTE
MANUFACTURED & GUARANTEED BY AMALGAMATED WIRELESS (A/SIA) LTD.

AWA Fisk Radiolette
(Empire State)
Australia

1934

1934

AWA Fisk Radiolette (Empire State)
& Cigarette Box (1936)
Australia

AWA
THE FISK
AWA
RADIOLETTE

1935 EKCO AC85
UK
Designer: Wells Coates

EKCO
MOULDINGS
ENQUIRIES ARE INVITED
FOR MOULDINGS IN ANY
SIZE, SHAPE OR COLOUR
ALL MOULDINGS SHOWN
ON THIS STAND ARE
THE PRODUCTS OF
EKCO BAKELITE PLANT
SOUTHEND-ON-SEA. ESSEX.
RADIO CABINETS
POWER UNIT CASES
RADIO COMPONENTS
ELECTRICAL FITTINGS

Mr. Rowe Mr. Hunt

Egypt. Mr. Hunt & Mr. Rowe.

SW86

EKCO AC86 1935

UK

Designer: Serge Chermayeff

VARIETY IN SPECIALLY STYLED FIXTURES

★ Specific styling of radio display fixtures, so that sets may be effectively displayed in various backgrounds in department stores particularly, has been deftly accomplished by International Radio Corp. of Ann Arbor, Mich., makers of Kadette sets.

Designed with taste and discernment, the fixtures are adaptable to gift shops and drug or jewelry stores, as well as sundry sections in department stores.

INTERNATIONAL RADIO
KADETTE 44 "Jewel"
USA

1935

AIR KING "Atlas"
and carrying case
USA

1935

EKCO AD36 **1935**
UK
Designer: Wells Coates

1935 RCA VICTOR Special Portable Phonograph
USA
Designer: John Vassos

RCA Victor Special

SPARTON 1186 "Nocturne"
USA
Designer: Walter Dorwin Teague

1935

1935 TRIUMPH "Midget" [aka RADIOGRAND]
Australia

RUBIS
Belgium

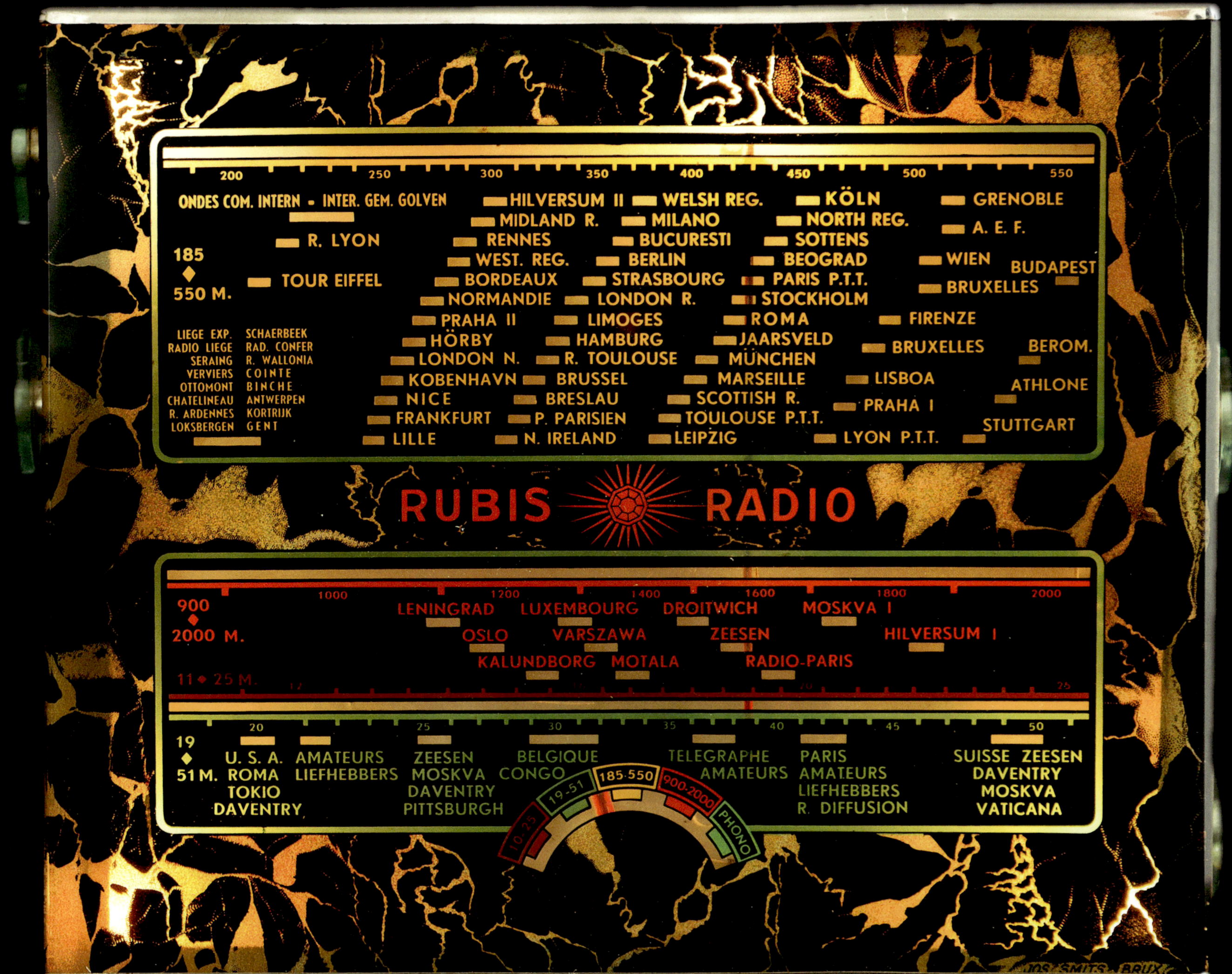

Colour in Radio
For the Nursery
For the Verandah
For the Library
The FISK
RADIOLETTE
Battery Operation

AWA Fisk Radiolette (F
Australia

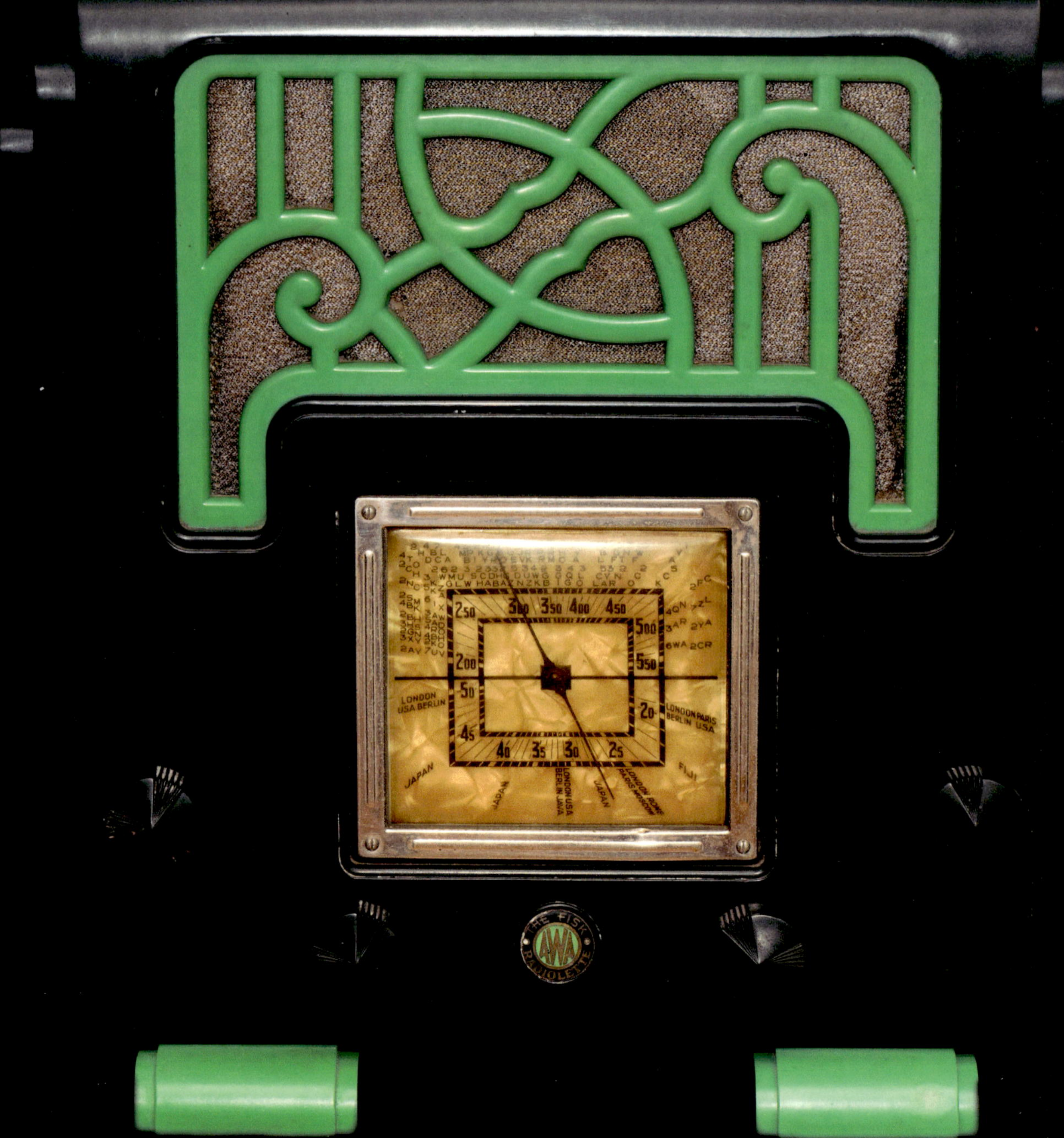
LONDON USA BERLIN
LONDON PARIS BERLIN USA
JAPAN
JAPAN
LONDON USA BERLIN JAVA
JAPAN
LONDON ROME PARIS MOSCOW
FIJI
THE FISK RADIOLETTE
AWA

RUBIS 57
Belgium

c1935

RADIO-GLO

RADIO-GLO
USA

1935

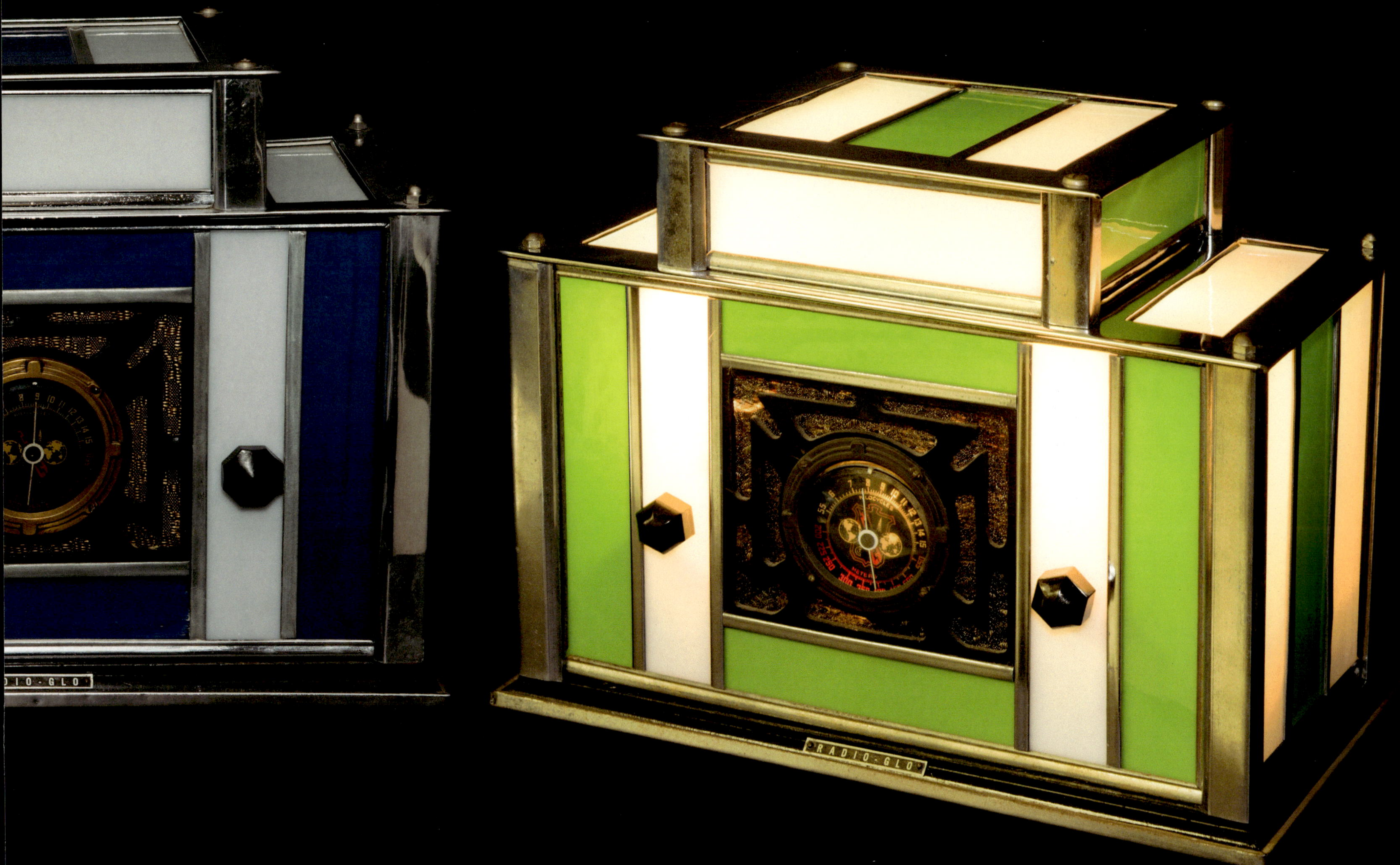

1936 RCA VICTOR 8T11 & 9K10

USA

Designer: John Vassos

EKCO AC97 1936
UK
Designer: Jesse Collins

SPARTON 566 "Bluebird"
USA
Designer: Walter Dorwin Teague

1936

Designer: Walter Dorwin Teague

RADIALVA TO-56 1936
France

1936 SPARTON 558 "Sled"

USA

Designer: Walter Dorwin Teague

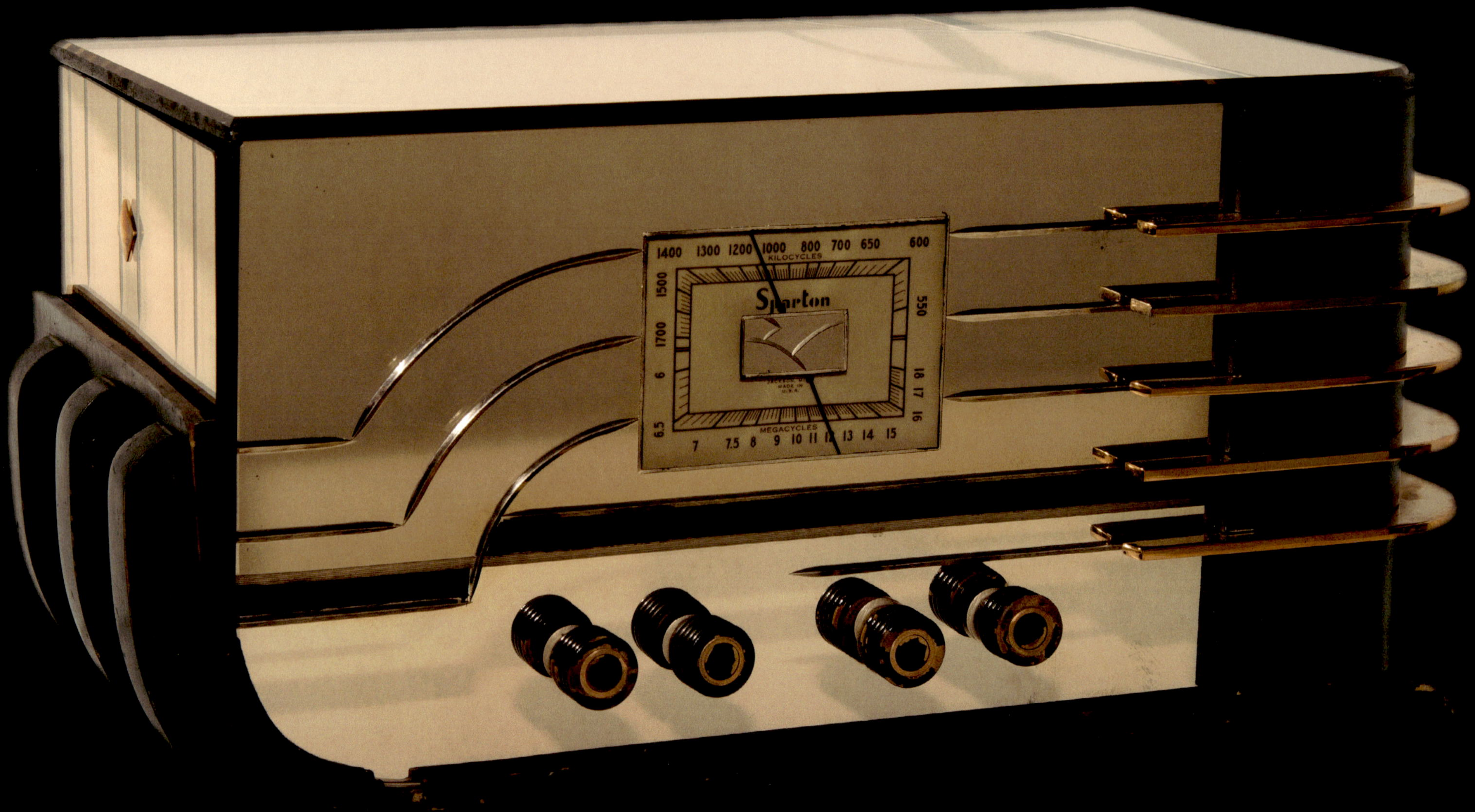

Designer: Walter Dorwin Teague

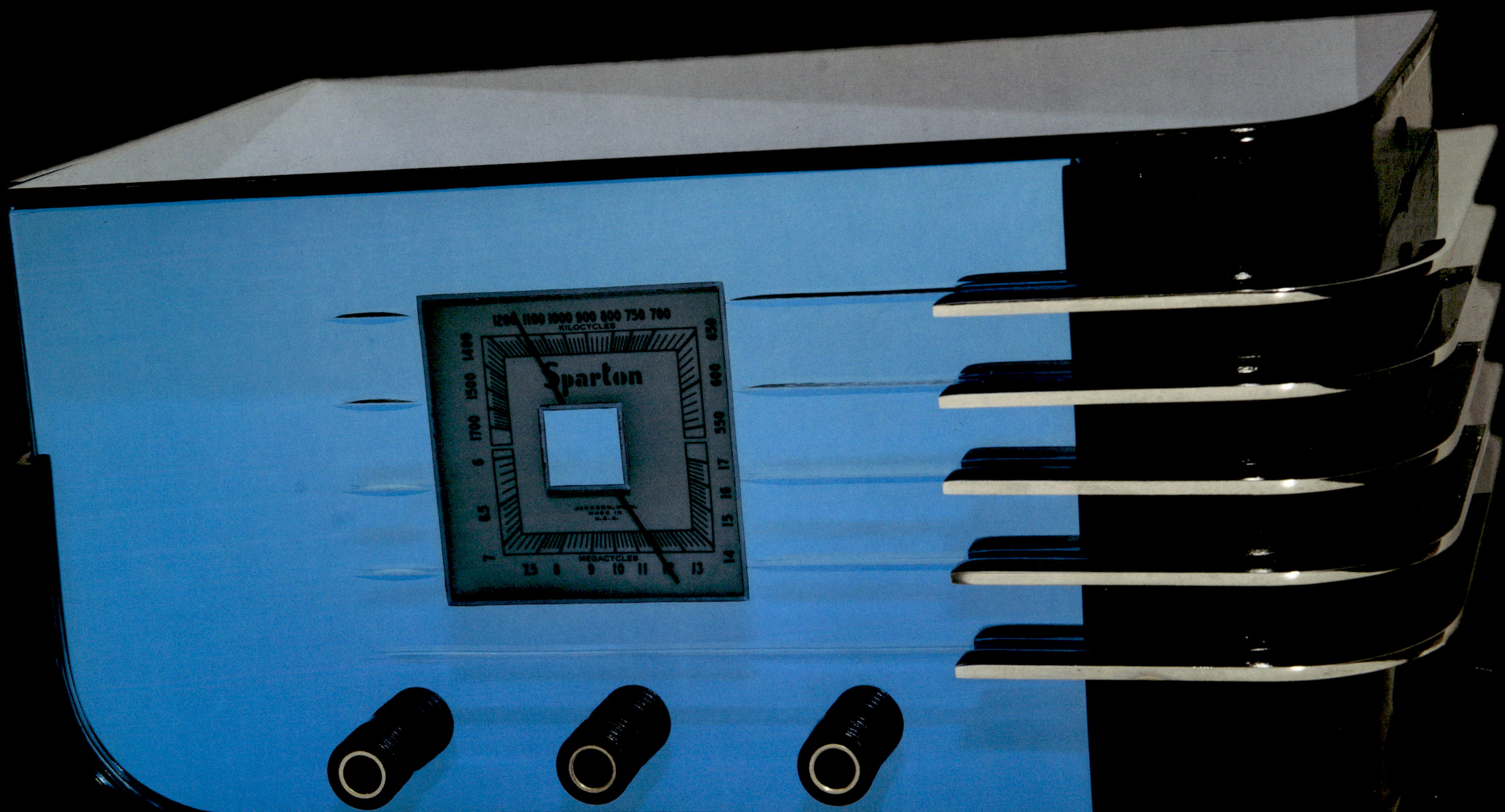

GREATER VALUE
FADA
Radio
The RADIO of TOMORROW
Today!
FADA
Radio
FADA The Radio of Tomorrow... TODAY
FADA RADIO
UNEXCELLED FOR SUPERB TONE

FADA 5F60 1936
USA

1935–1938 REMLER 46, 26, 40 "Scottie" USA

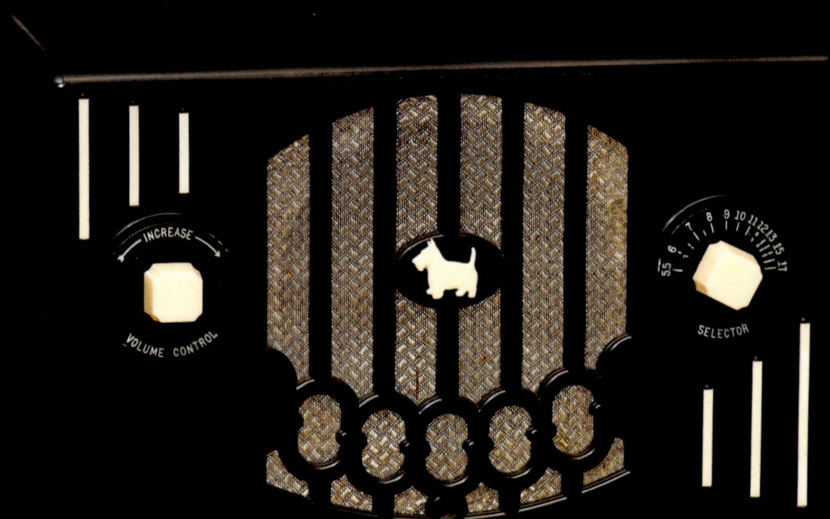

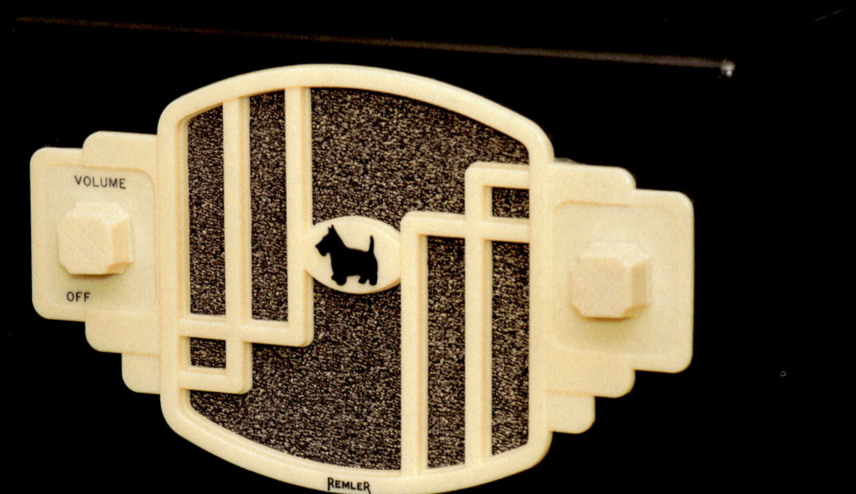

REMLER 40 with Blue Mirror Case 1938
USA

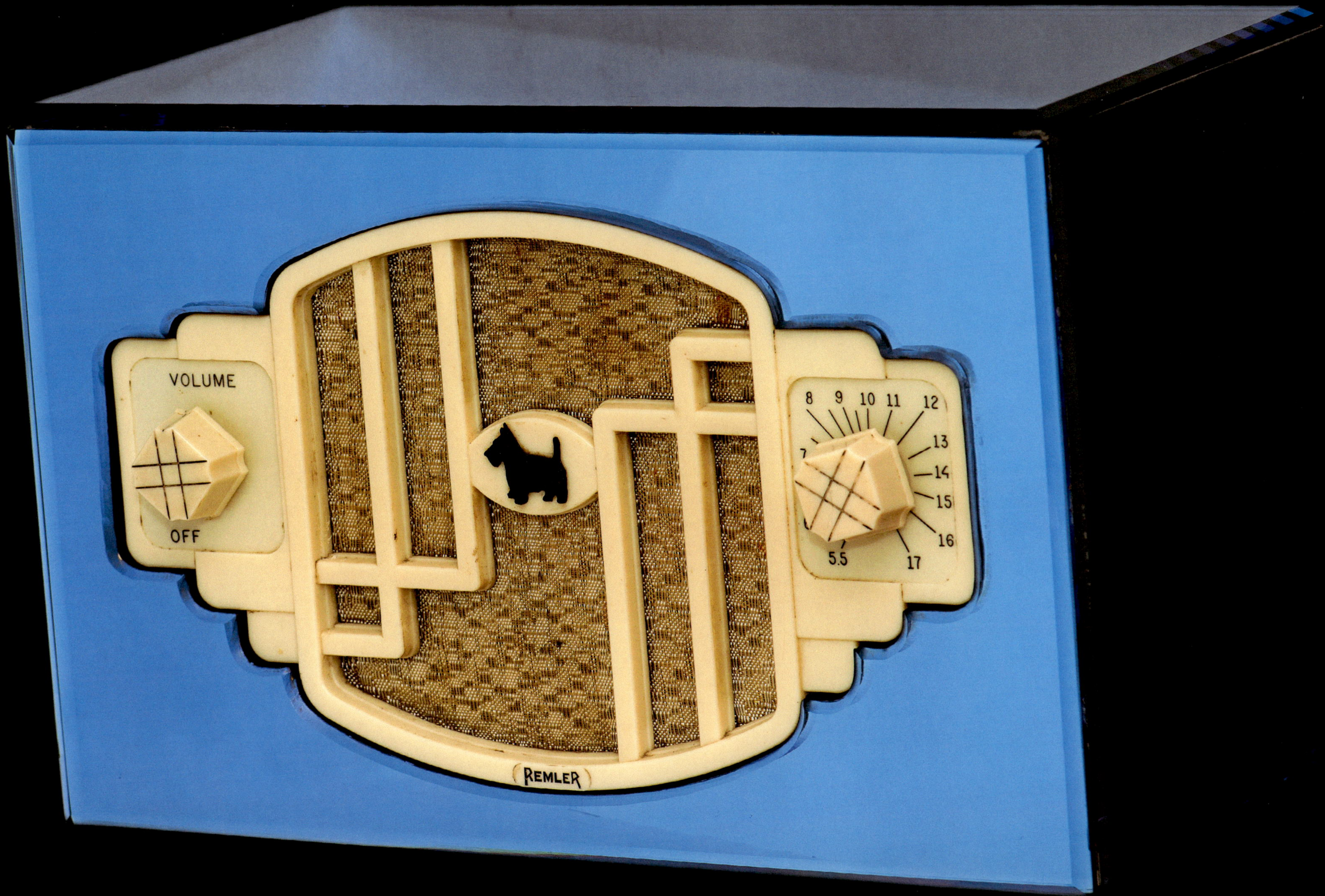

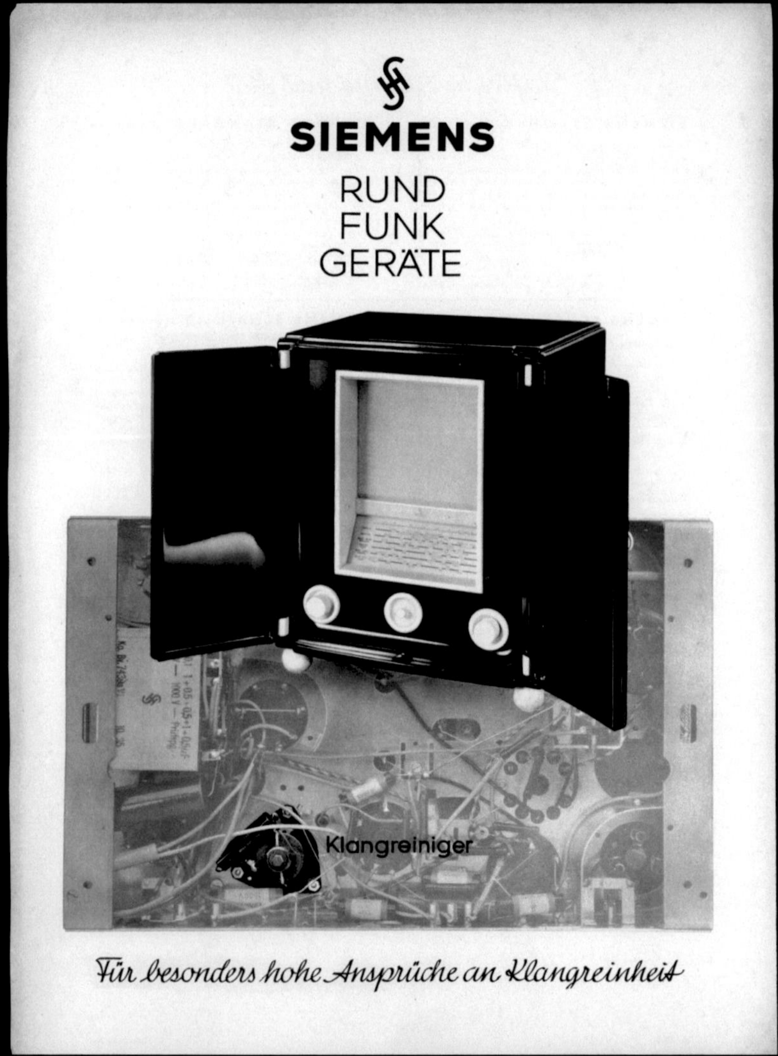
SIEMENS
RUND
FUNK
GERÄTE
Klangreiniger
Für besonders hohe Ansprüche an Klangreinheit

1936 PHILCO 444 (People's Radio)
UK

EMERSON AU-190
USA

1937

1937 WELDONETTE
Australia

KADETTE "Classic"
USA

193

KADETTE
KADETTE
RADIO
The World's Best Value
KADETTE

KADETTE K25 "Clockette" 1937
USA

Emerson
70 80 100 120
60 150
55 170
175 550
200 500
250 300 400
METERS
Emerson Radio & Phonograph Corporation N.Y.U.S.A.

COLLEGE COLORS, too:
Maroon & White
Black & Orange
Blue & White
A BEDROOM "Symphony"
Selection of Six Gem-like Color Combinations
Model AX-235—5-Tube AC-DC Superhet. ● Amer. Broadcasts and Police Calls ● Electro Dynamic Speaker ● Automatic Vol. Control ● Built-in Antenna ● Also available in Natural Onyx, Brazilian Onyx, Tyrolean Red
$17.95
SNOW WHITE MODEL BM-247
Your CHILDREN'S Room
Snow White and Seven Dwarfs
5 Tubes (incl ballast tube) ● AC-DC American Broadcasts, Police ● Dynamic Speaker ● Automatic Overload Control ● Built-in Antenna.
$14.95
"Emersonize"
YOUR LIVING ROOM, KITCHEN, BEDROOM, PLAYROOM, OFFICE
Any EMERSON is an IDEAL GIFT
With the variety of beautiful STYLES, with the popular price range—and with an Emerson for every purpose and every purse — you can, and SHOULD get this greater enjoyment of radio programs.
Only Emerson has the "Miracle Tone Chamber."
Come in and see how NATURALLY Emerson Radio fits into any decoration and utility scheme—how easy it is to buy!

EMERSON AX-235
(Little Miracle)
USA

1937

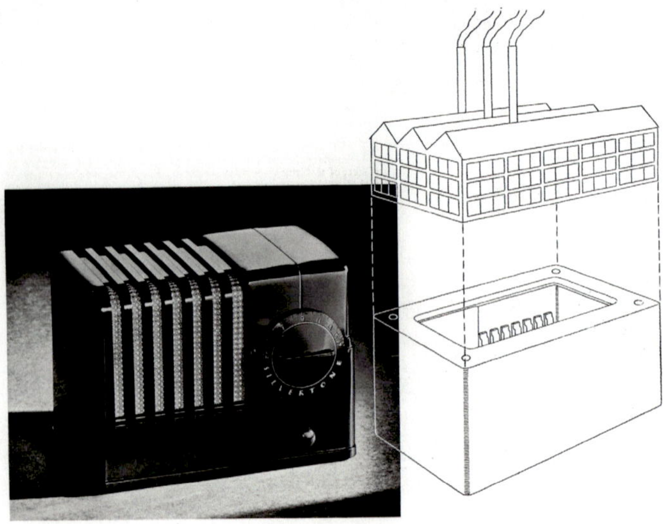

A FACTORY IN A FOOT OF SPACE

Behold a mold for making a radio cabinet of Durez. An entire factory in itself. From it comes the finished case ready for assembly . . . lustrous, sculptured, complete. Produced in a matter of minutes.

Where else can you duplicate such saving of time and labor and space? Or achieve such beauty of results?

Your product may be large or small, intricate or simple. Durez will give it new sales appeal. Make it lighter in weight. Give it a touch and feel that's almost human. Lend it lustrous color and finish that will not wear.

All in one simple, almost instantaneous operation!

And it goes without saying that Durez is one of the strongest of modern materials. Resistant to heat, moisture, acids. Will not chip nor peel. Self-insulating, non-resonant.

We will be glad to send interested executives interesting suggestions for using Durez, the hot-press phenolic molding compound. Just write General Plastics, Incorporated, 110 Walck Road, North Tonawanda, New York.

"OUR SILVERTONE HAS BEEN OUR FASTEST-SELLING RADIO," says Sears-Roebuck executive.

"When the Durez cabinet comes out of the press it's virtually finished, ready to be dropped over the chassis and sold," says D. M. Nelson, Vice-Pres., in charge of merchandising. "Because so many finishing operations are eliminated, these Durez radio cabinets are quite economical to produce — and that's a big factor in giving so much value for only $10.95."

GENERAL PLASTICS'

DUREZ

ONE OF INDUSTRY'S MOST VERSATILE MATERIALS

• 175 •

ZENITH "Radio Nurse" & "Guardian Ear" 1937
USA
Designer: Isamu Noguchi

1937 SPARTON "Polo Club"
USA

CENTRAL DES ONDES
France

UNKNOWN MAKER
Belgium

c1937

c1937 UNKNOWN MAKER
Belgium

MAGNADYNE "Radio Balilla" 1937
Italy

Graduation Present

Choice of 5 Color Combinations

Model BM-258 — 5 Tubes (including ballast tube) AC-DC
- American Broadcasts, Police Calls
- Dynamic Speaker
- Automatic Overload Control
- Built-in Antenna
- Plastic cabinets.

$14.95

1939 EMERSON MODELS

EMERSON BM-258
(Big Miracle)
USA

1938 EMRIC 638A "L'emric"
France

1937 ZENITH 5S218 (Cube)
USA

STEWART WARNER 91-513 (Spade) **1939**
USA

DETROLA 105 "Faust" **1937**
USA

NEW S5F60 FADALETTE SERIES

The 5F60—5 Tube Superheterodyne Chassis, already famous for its fine tone and superior performance in sensitivity and selectivity, is now available in the same series of Catalin cabinets in seven color combinations, already announced for the 5F50 Catalin series.

Model S5F60AG illustrated above and the other color combinations and chassis specifications are listed on the reverse side of this sheet.

$19.95

FADA 53
USA

1938

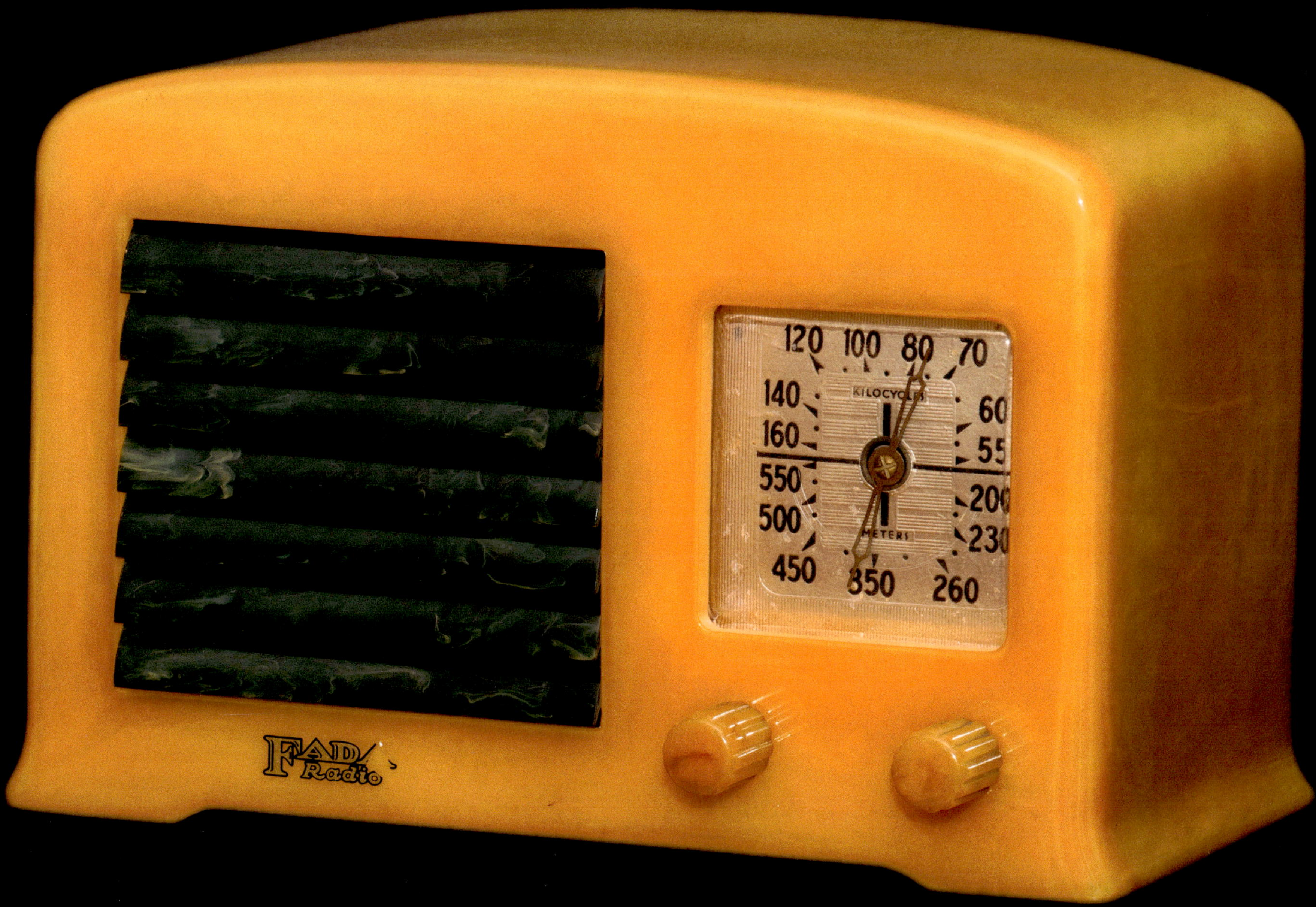

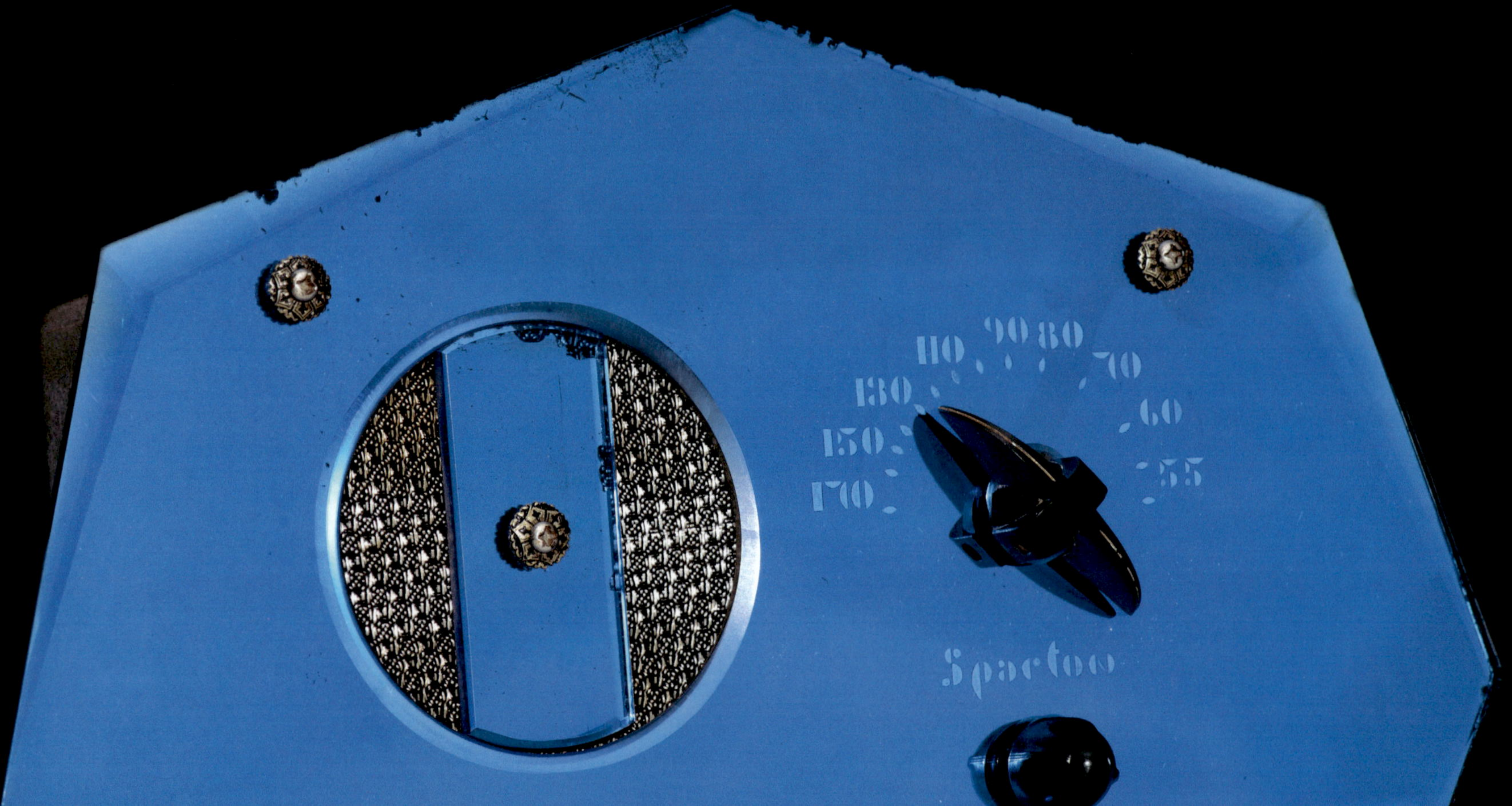

Created By the Su
Hand of Experienc

Dealers everywhere have asked for this new version of Sparton, wrought in Midnight Blue Mirror Glass like its amazingly successful predecessor — the Sparton Bluebird model. And now it is ready for Christmas selling — a merchandising triumph that will sell at an accelerated pace as thousands are given the opportunity to appraise and appreciate its beauty and surpassing performance.

BLU
Per

"CYNOSURE"

Style Leader of All Radios

Here is beveled Midnight Blue Mirror Glass modeled in a rhythmic, modern design enriched with touches of silver glints and ebony black. It is a magnificent, artistic coalition of contour and colors . . . a worthy successor to the widely accepted Sparton Bluebird Model . . . a radio thoughtfully designed to harmonize with all interiors, whether modern or traditional. Truly an original and imaginative gift whose beauty and character are too elusive to be trapped in words. It is featureful, too, with 5-tube (with ballast) AC-DC chassis and rich-tuned dynamic speaker.

FREE DISPLAY

Pric
Gift

The stunningly beautiful display unit shown above is given FREE to all Sparton dealers. The display lends itself to dramatic use in show window or store, and it has the power of creating sales because it glorifies the smartest radio yet to reach the market.

Ready Now For Immediate

THE SPARKS-WITHINGTON CO

SPARTON 409-GL (7-sided)
USA
Designer: Walter Dorwin Teague

1938

1938 GENALEX
Australia

HALSON A5 "Dwarf" 1938
USA

GET THIS TODAY!

The Full Story of

Halson's

amazing new miniature

The

"DWARF"

Fits into the palm of your hand

• Here is a brilliant performer and superbly beautiful little set that is a real PROFIT LEADER — not a *loss* leader.

It is the type of set that will lead your customers away from the snare of cheapness by giving them a far better set for only a trifle more.

5-tube AC-DC . . . Large dynamic speaker . . . Marvelous tone . . . Broadcast and police calls . . . Exclusive new onyx-like cabinets . . . Red, green or ivory, all priced the same!

$15.00 LIST

You can make money and hold the good will of your customers, by selling "The Dwarf"!

Get the whole story of the latest Halson hit and five other smart table models. Send *today*—and handle the line that's on its way to record popularity.

HALSON

RADIO & TELEVISION CORP.
MERIDEN, CONN.

ANOTHER SILVERTONE STYLE SENSATION!

IT'S NEW · · · IT'S MODERN · · · IT'S 100%
INSTANT PUSH-BUTTON TUNING

4-STAR FEATURE

Rear View

Its striking design will convince you it is the only small radio of its kind!

$12.95 CASH

5 tubes . . . automatic push-button tuning . . . 6 push buttons . . . complete list of call letters supplied . . . tuning range: 545 to 1720 K.C., including some police calls . . . ball bearing condenser . . . large ball-type tuning knob . . . 5-inch full floating dynamic speaker . . . chassis completely housed in molded cabinet . . . back of cabinet as attractive in appearance as front . . . attached 20-ft. aerial included. A real four-star feature.

This new Silvertone "Rocket" Compact is destined to zoom to popular favor with even greater success than our sensational Streamliner shown at the right-hand bottom of this page, which two years ago made radio history and won first prize in the Plastics Exhibition! Sears beautiful Silvertone "Rocket" compact was designed for modern people living in a modern world. Combines breath-taking beauty of line with truly remarkable performance! In addition, it offers you the very latest of modern features.

It's a full 5-tube radio!

Every tube in this remarkable little radio is a working tube! No dummy or "ballast" tubes! Its large, full-toned dynamic speaker provides rich, clear tone quality . . . ample volume, more than is usually obtained from much larger radios. Attached aerial. The ideal radio for bedroom, den or any other room where a small set is needed. Grand for an end table or near your favorite chair. Cabinet is polished black Bakelite that will match any style of room decoration. Size: 6⅛x6⅕x11¾ in. long.

Instant push-button tuning!

Six instant action push buttons are built into the top of the "Rocket." These can be pushed from any angle, sitting or standing. So simple a child can operate them with ease. Large ball-shaped tuning knob fits the hand perfectly. From every angle this radio presents a pleasing appearance, the rear being as attractive as the front. For 110-volt 60-cycle A.C. If for D.C. or 25-cycle operation so state and send $1.50 extra. *Mailable.* Shipping weight, 13 lbs.
57 K 6110—Complete with tubes. $12.95

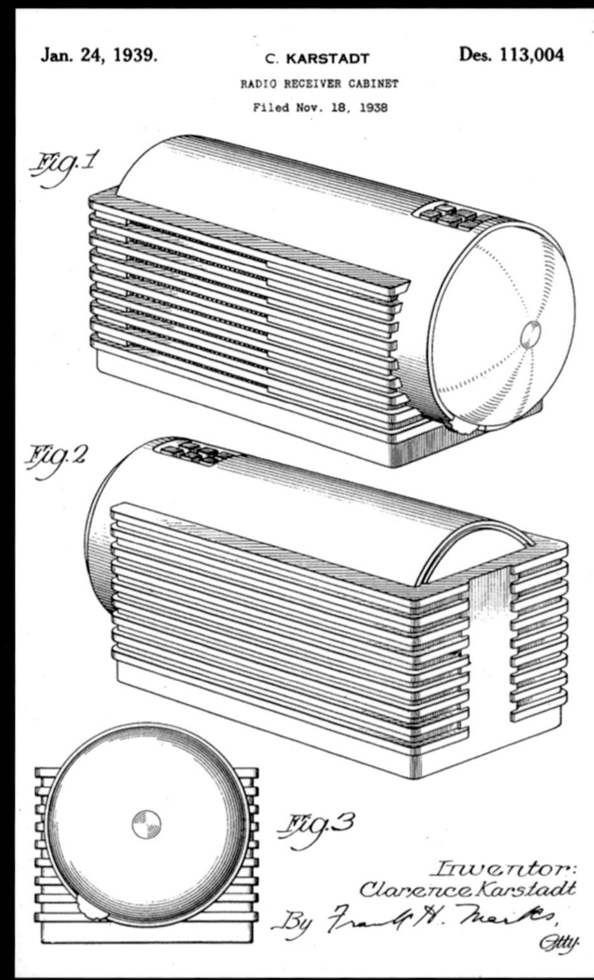

SILVERTONE 6110 (Rocket)
USA
Designer: Clarence Karstadt

1938

New 1941 Models

ARVIN

Radios

THE FAST-SELLING LOW-PRICED LINE

16 Hot Numbers to Sell

Beautiful Onyx Catalin

5-TUBE AC-DC SUPERHET

Looks like a Jewel

MODEL 532A

ARVIN

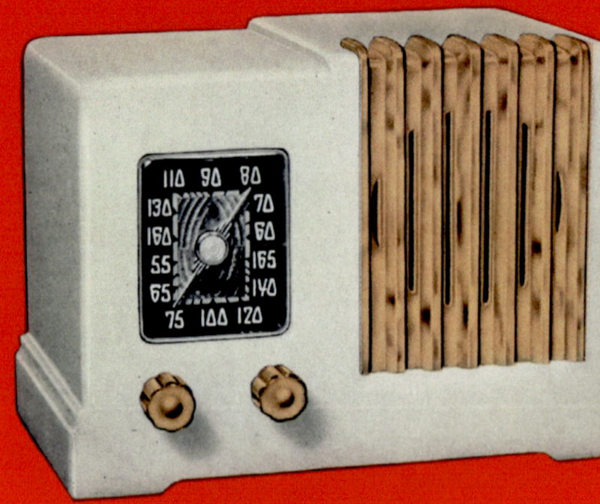

Built-in Loop Antenna

Cabinet Size: 8¼″ wide, 6″ high, 5¼″ deep

HERE are the most glamorous radios you have ever seen—in catalin, the jewel of plastics—with 5-tube superhet performance that matches their jewel-like beauty.

The two models shown on this page have the same features and specifications. They differ only in the color of the cabinets as illustrated.

Beautiful beyond description, these models will pep up the sales of any dealer who shows them. Priced low for quick turnover.

These models sell on appearance as well as performance. Here are the features that make them such fine radios:

5-Tube Superheterodynes using these tubes: 12SA7GT, 12SK7GT, 12SQ7GT, 50L6GT, 35Z5GT, including power rectifier. Operates on AC or DC.

Airplane Type Lighted Dial has double-ended cameo pointer with fluorescent plastic insert for easy reading.

Two Dual-Purpose Tubes give performance equivalent to a 7-tube radio.

Quadruple Tuned I.F. Amplifier gives knife-edge selectivity.

Loop Antenna built-in, with connection for external aerial if desired.

Band Coverage 540 to 1750 kc.

Electrodynamic Speaker for fine tone quality.

Rich Burgundy Catalin

5-TUBE AC-DC SUPERHET

It's a Beauty!

MODEL 532

ARVIN

Built-in Loop Antenna

ARVIN 532
USA

1938

Fast-Selling Radios

130
70
160
60
55
170
65
ARVIN
75
95
120

1938

CROSLEY G1465 (Split-Grill)
USA

HMV (Tower)
Australia

1938

1938 AIRZONE 537 "Symphony Leader"
Australia

FADA 52
USA

1938

Emerson
Radio and Television

EMERSON BT245
USA

1938

1938 DETROLA 219 "Super Pee-Wee Midget"
USA
Designer: George W. Walker

DETROLA "Pee-Wee Midget"
USA
Designer: George W. Walker

1938

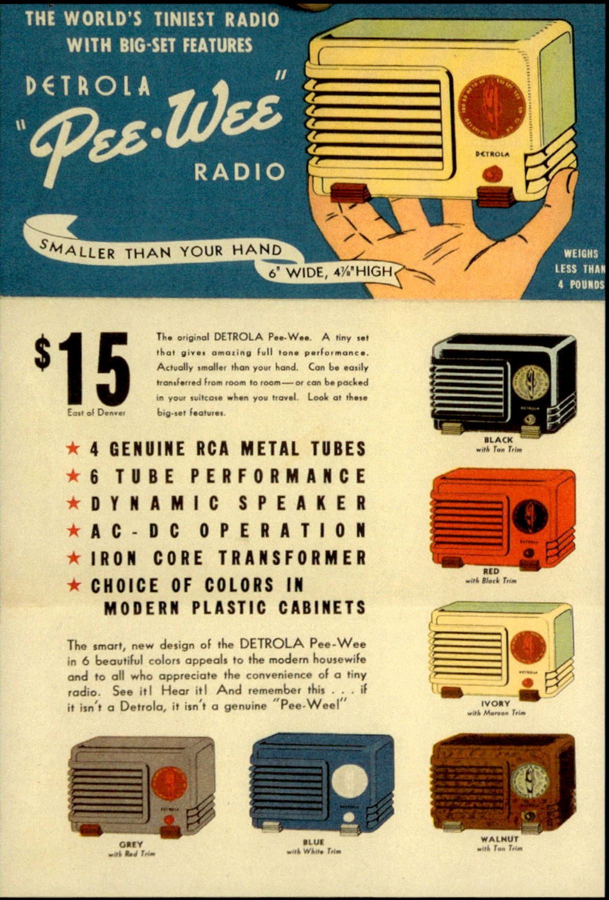

1938 DeWALD A501 (Lyre)
USA

MULLARD Master 50
Australia

1938

TOM THUMB
TOM THUMB
TOM THUMB
TOM THUMB
TOM THUMB

AUTOMATIC TOM THUMB 995
(Non-Deco or Oval Dial)
USA
Designer: J. Samson Spencer

1938

c1938 ALFA "Cyclope Studio"
Belgium

STEWART WARNER "Varsity"
USA

1938

938 BRITON [PHILIPS] "Theatrette"
Australia

EMERSON BD-197 (Mae West) 1938

USA

Designer: Alexis Sakhnoffsky

1938 AWA Radiola (Egg Crate)
Australia

CORD "Zephyr"
USA

c1939

130 100 80 70
60
55
175
200
400 300 250
METERS
FADA Radio

FADA L56
USA
c1939

80 90 110
70 130
60 150
55 170
KILOCYCLES
Motorola
AMERICA'S FINEST RADIO FOR YOUR CAR - FOR YOUR HOME
Motorola
RADIO
GALVIN MFG. CORPORATION
4545 AUGUSTA BLVD. - CHICAGO, ILL.

Motorola
America's finest
ROOM RADIOS
Motorola
AMERICA'S FINEST RADIO
Motorola
AMERICA'S FINEST RADIO

MOTOROLA 52 (Vertical Grill)
USA

1939

1939 RCA Victor RC-350
"Tulip Grill"
USA

JANUARY, 1939

LISTEN

RCA

A 4-Page Advertisement of the Radio Corporation of America No. 18

To those who must endure long years of illness, religious programs by radio bring consolation and encouragement. Shown is Mrs. Martin J. Dufficy of Washington, D. C. who has been confined to her bed for 14 years. Says Mrs. Dufficy: "Radio messages from religious leaders have lightened the crosses of thousands and made them easy to bear."

Religion and Radio

Religious freedom is one of the foundation stones of the United States. Hence the broadcasting of religious programs is a vital public service. First principle is that time devoted to religion is donated. The reason is obvious. To sell time would give an advantage to the religious organization with the most funds available for such use.

In general, NBC relies for religious programs on the central or national agencies of the great religious faiths. Many other groups are also given time on the air on special occasions.

NBC serves religion in many ways. First, it serves the church, in the sense of its service to the chief faiths. Second, it serves individual listeners regardless of whether they are, or are not, members of any faith. Third, NBC religious programs further mutual respect and understanding by making it possible for adherents of one faith to hear the views of another. Yet no speaker ever attacks another faith, nor seeks to change the religious convictions of listeners. As a further service, radio brings to millions special religious events which previously were heard only by thousands.

Commenting on radio and religion a speaker recently said: "I can think of no greater benefit that can come to mankind at this time than the continued preaching of the need for understanding and for tolerance among the different peoples, races and creeds of the world. If religion can carry that message to the hearts and to the minds of all the people throughout the world, radio will have justified itself a thousand times and more, for, after all, man's hope and aspiration for the last two thousand years has yet fully to be realized —Peace on earth, Good will to men."

RADIO'S SERVICE TO RELIGION

SEE PAGES 2 and 3

RCA Victor "Little Nipper II" 9TX
USA
Designer: John Vassos

1939

BP "Mickey Mouse"
Australia

SYMPHONY (Split Grill)
USA

1939

1939 ASTOR "Mickey"
Australia

MAJESTIC 5T-0
USA

1939

1939 KADETTE L-25 "Topper"
USA

UNIVERSAL c1939
USA

Sparton
160 140 120 100 80 70 60 55
Sparton

SPARTON 500C "Cloisonne"
USA
Designer: Walter Dorwin Teague

1939

KILOCYCLES
DETROLA
METERS
80 100 120
70 130
60 150
55 170
200 500
230 460
250 300 375 430

DETROLA 281 (Split Grill)
USA

1939

LOCYCLES 55 60 70 80 90 110 130 150 170 POLICE
DEWALD

SENTINEL 195 ULTA (Push-Button)
USA

1939

1939 SENTINEL 177U
USA

FADA 188 "All American"
USA

1939

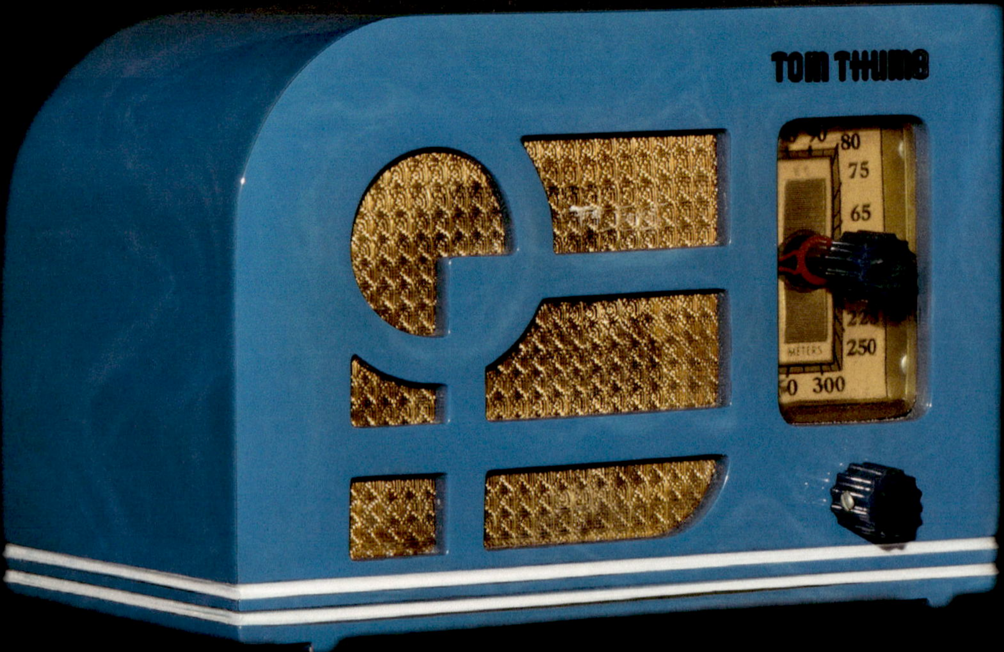
TOM THUMB

TOM THUMB

TOM THUMB

AUTOMATIC TOM THUMB (Deco) 1939
USA
Designer: J. Samson Spencer

c1939 RADIO LAMP
USA

GAROD 1450 (Peak Top) **1939**
USA

939 PHILCO "Transitone" & SIMPLEX UA52P
USA

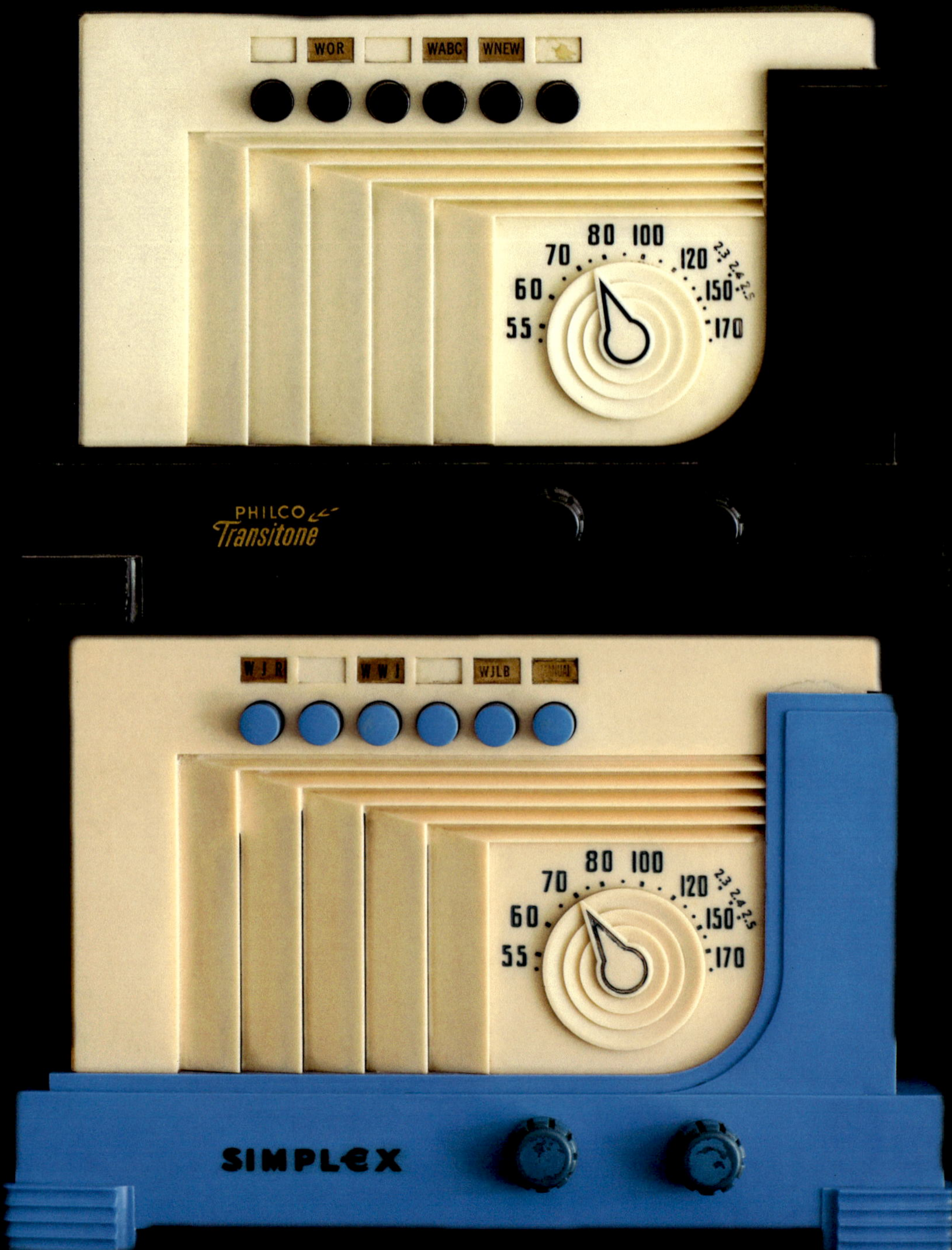

RCA VICTOR 96X c1939
USA
Designer: John Vassos

1940s

USA

USA

1940s

Sentinel
55 60 70 80 90 100 120 140 160
PHILCO
55 60 70 80 90 110 130 150 170

1940s

USA

DENMARK

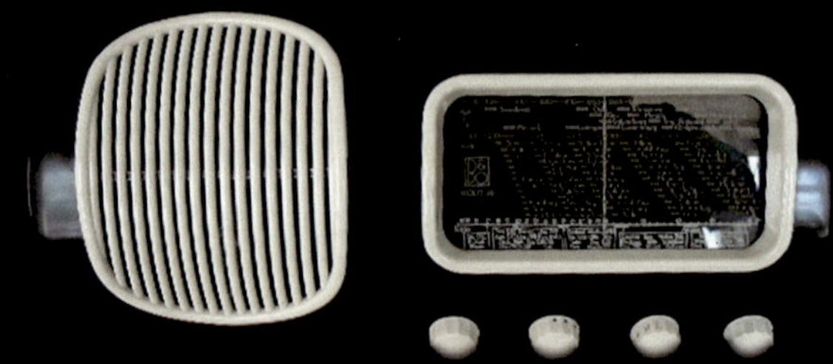

AUSTRALIA

SPAIN

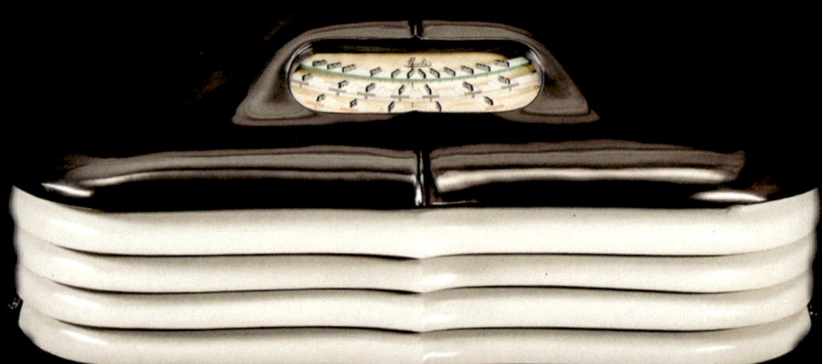

1940s

UK

CANADA

ITALY

CZECHOSLOVAKIA

FRANCE

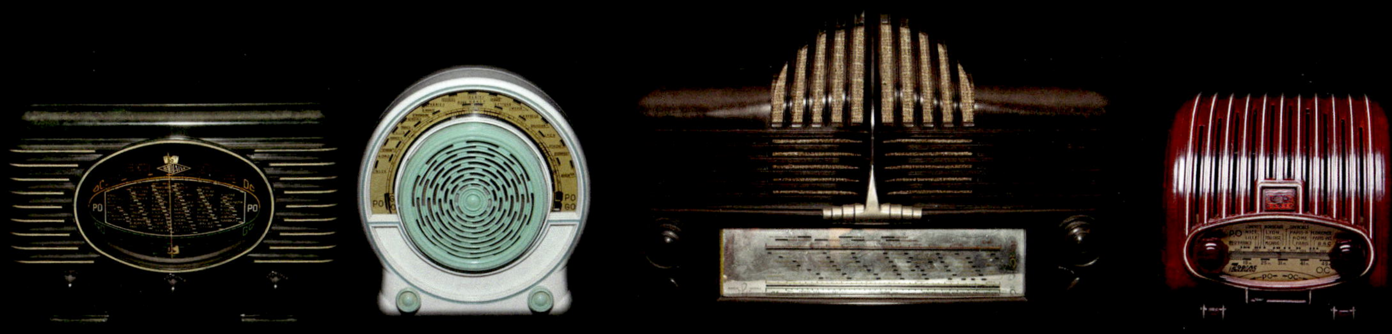

1940 ESPEY "Midget"
USA

EGM
Mexico 1940

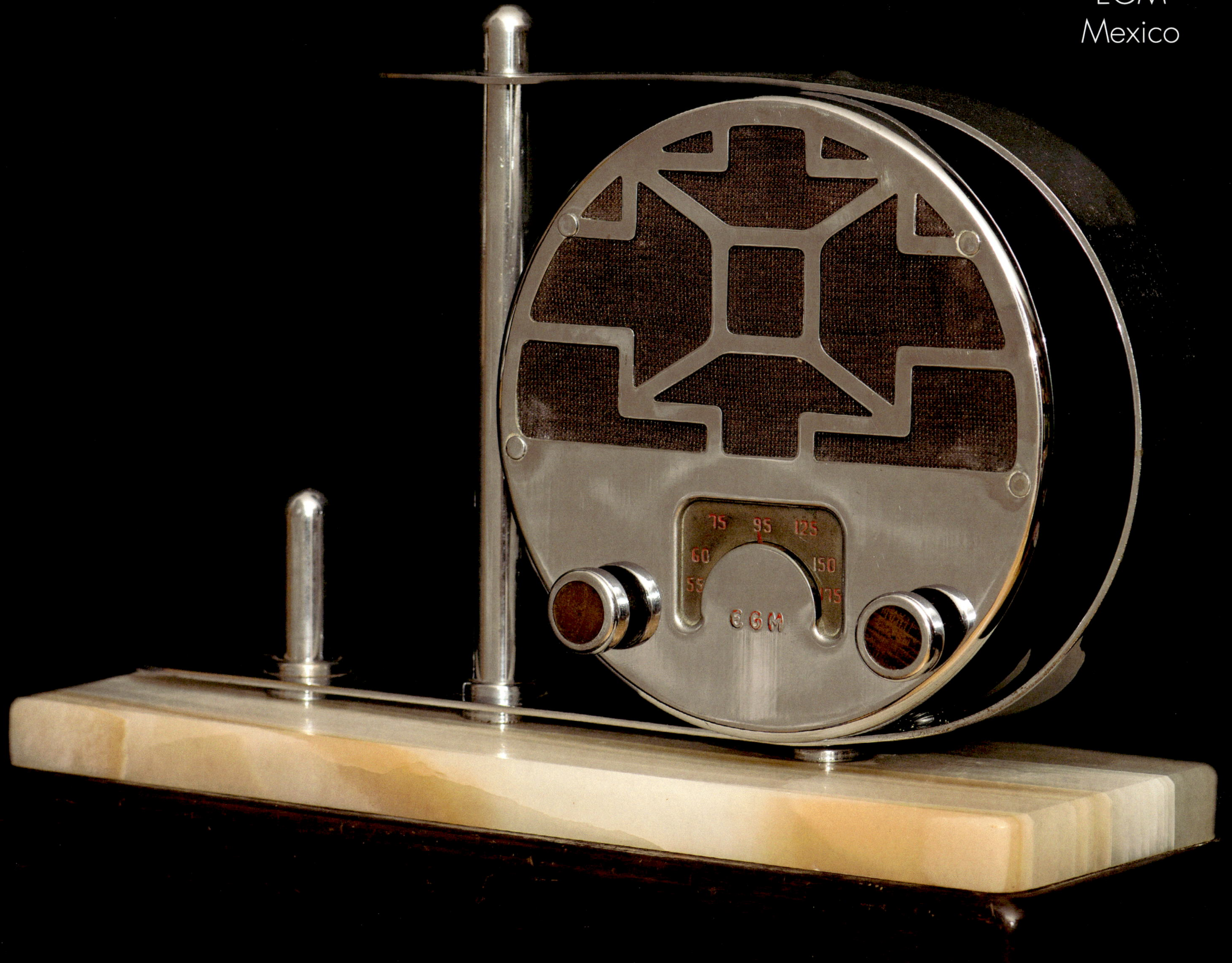

Lad hendes Juledrøm gaa i Opfyldelse
B&O
BEOLIT 39
Glæd hende med B. & O.'s nye Strømlinie-Super-Beolit 39 — hun har straks set, hvor elegant og raffineret den virker i det fornemme sorte Kabinet af Beolit med Rammer, Knapper og Gitter i Elfenbensfarve. — Modtageren er B. & O.'s effektive Super med 3 Bølgeomraader, Foto, Plan-skala, Lyndrev, Tonekontrol, aut. Fadingkon-trol og permadynamisk Koncerthøjttaler. — Ga hen og hør den hos vor Forhandler. Prisen e — inclusive 4 Rør — kun Kr. 275.
BANG & OLUFSEN A/S — STRUER

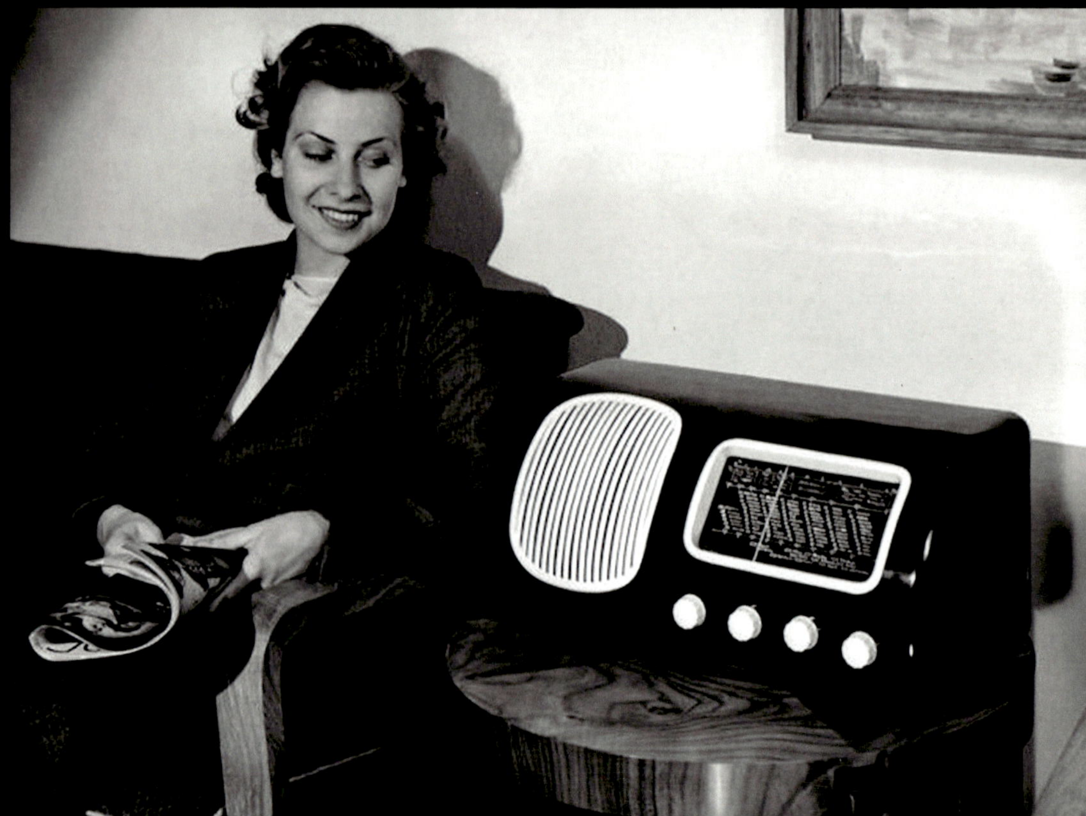

BANG & OLUFSEN "Beolit 40" Denmark

1940

1940 STEWART WARNER "Good Companion"
USA

65 70 90 110 130
60
55
185
200
230 250 400 450
150
160
550
500
KILOCYCLES
METERS
Lafayette

MEDIE
CORTE
PHONOLA

PHONOLA 573 c1940

Italy

Designer: Livio Castiglioni

Addison
"5" now in Catalin
The popular Addison Model "5" is now coming down the assembly line in new "eye-catching" Catalin cabinets—mottled green, mottled maroon and mottled cream—all at the same prices as the wooden cabinet models.
Now six cabinets to choose from in the Addison Model "5" — Mahogany, walnut, bleached birch and the three new Catalins.
We expect to start shipping them to our dealers from coast to coast this month.
Addison RADIO
Addison Industries Limited
MANUFACTURERS OF NORGE AND ADDISON PRODUCTS
A. CROSS & COMPANY LIMITED
NATIONAL FACTORY SALES & SERVICE
WINNIPEG
TORONTO
MONTREAL
Radio Trade-Builder—March, 1946

ADDISON A5 (Court House)
Canada

1940

1940 GAROD 126 (3 Ring)
USA

DUCATI RR3404 "Paniere" **1940**

Italy

1940 RADIALVA "Super-Groom"
France

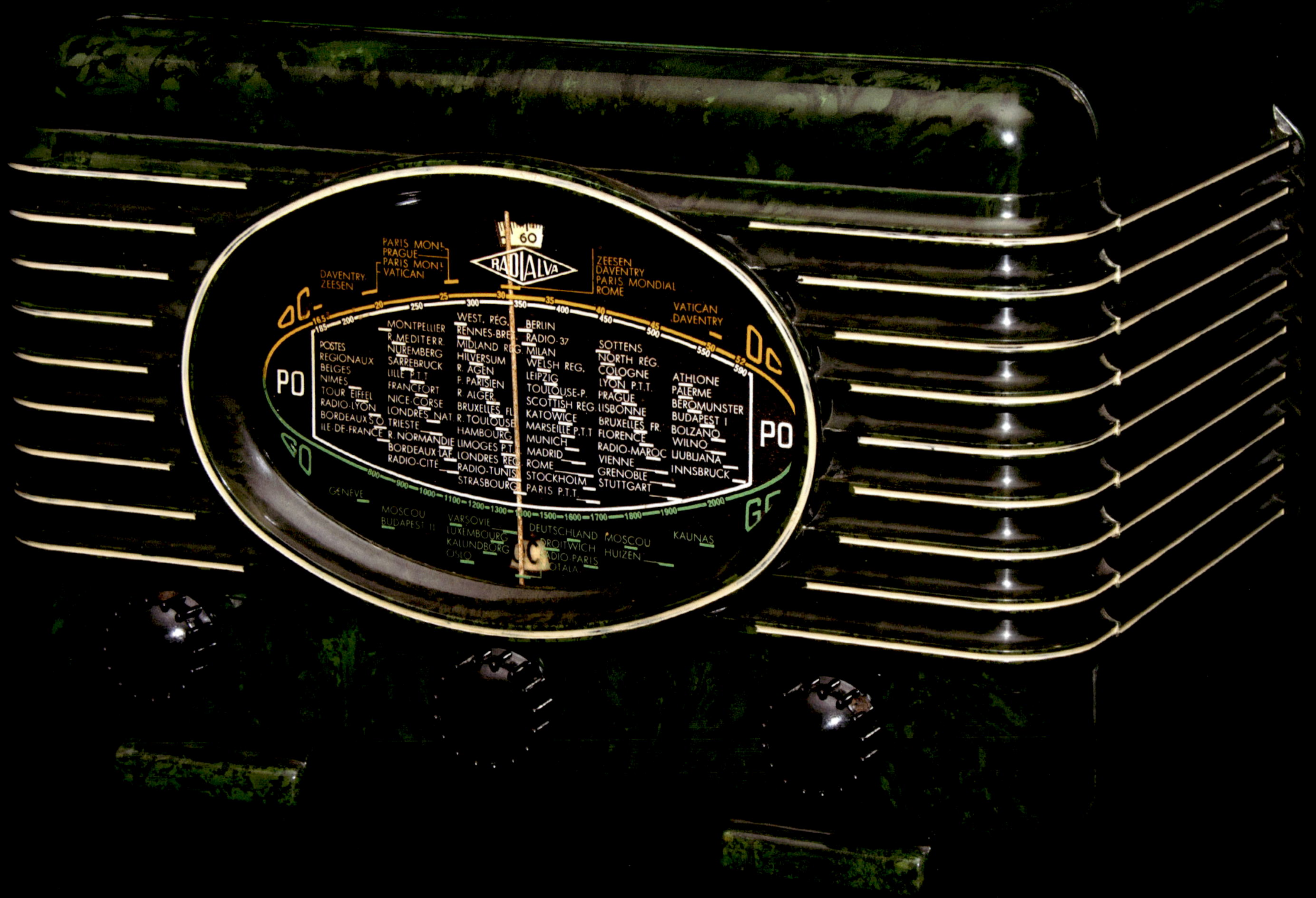

ADDISON A2
Canada

1940

1940

EMERSON 564
(Slot Grill)
USA

SILVERTONE "Commentator" c1940
USA
Designer: John R. Morgan

80 90 110
65
60
55
130
150
170
Motorola
80 90 110
65
60
55
130
150
170
Motorola

MOTOROLA 50XC (Circle Grill) 1940
USA

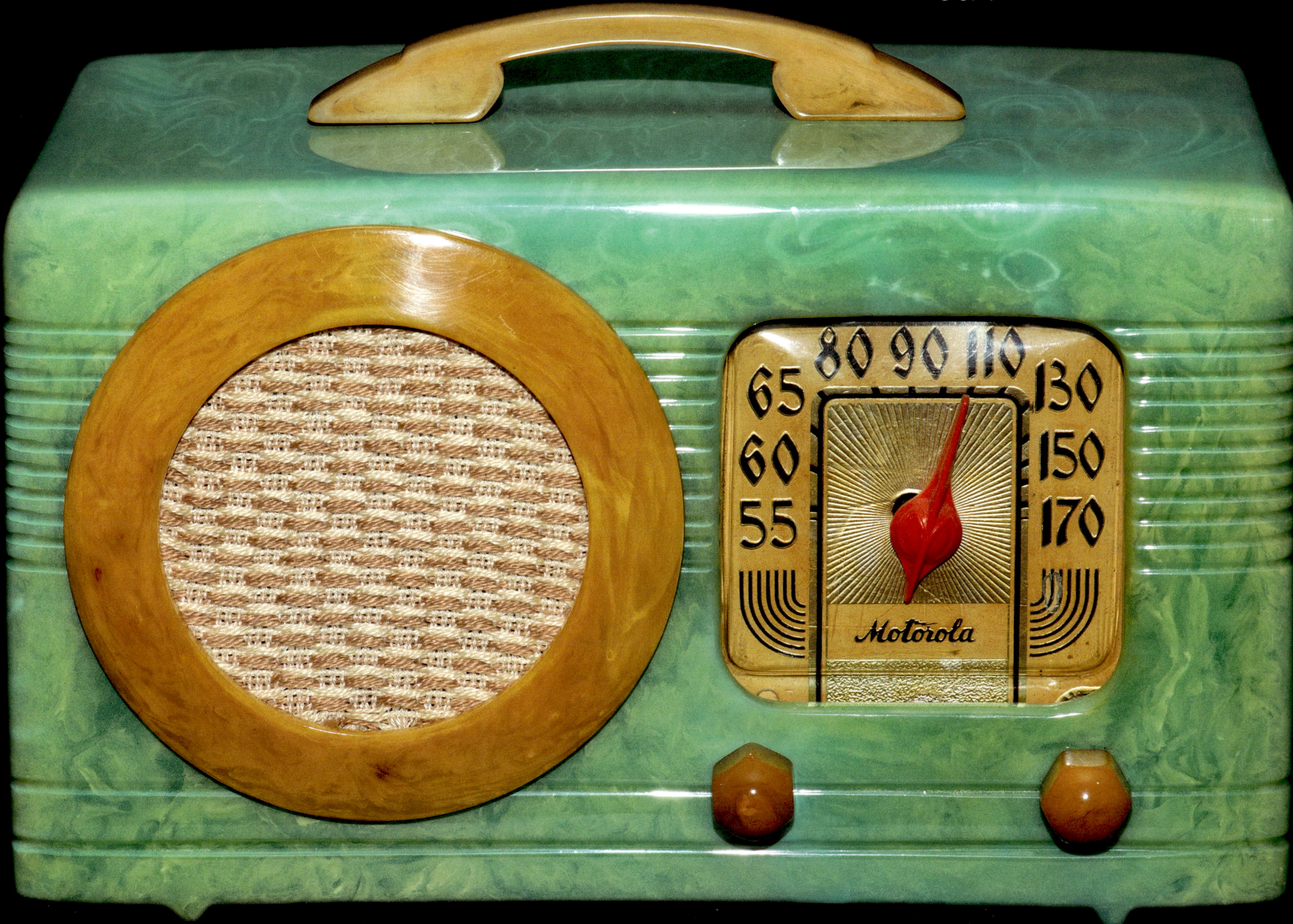

1941 IBERIA 4153
Spain

GENERAL ELECTRIC L570
USA

1941

1941 GENERAL ELECTRIC L-622
(Jewel Box)
USA

EMERSON EP375 (5+1) USA

1941

Emerson Radio and Phonograph Corp. N.Y.U.S.A.

EMERSON 400 "Patriot" 1941
USA
Designer: Norman Bel Geddes

4 Red, white or blue Emerson portable is housed in an Opalon cabinet fabricated by Ortho Plastic Novelties, Inc. Grille, knobs and pointer of cellulose acetate in contrasting colors injection molded by Worcester Molded Plastics, Inc. Such applications may be somewhat limited until defense needs are met

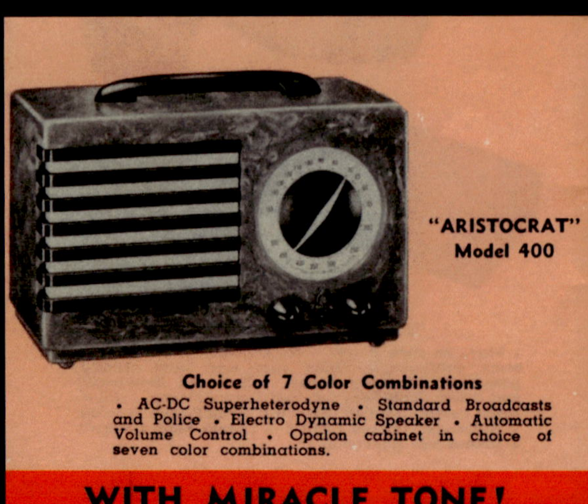

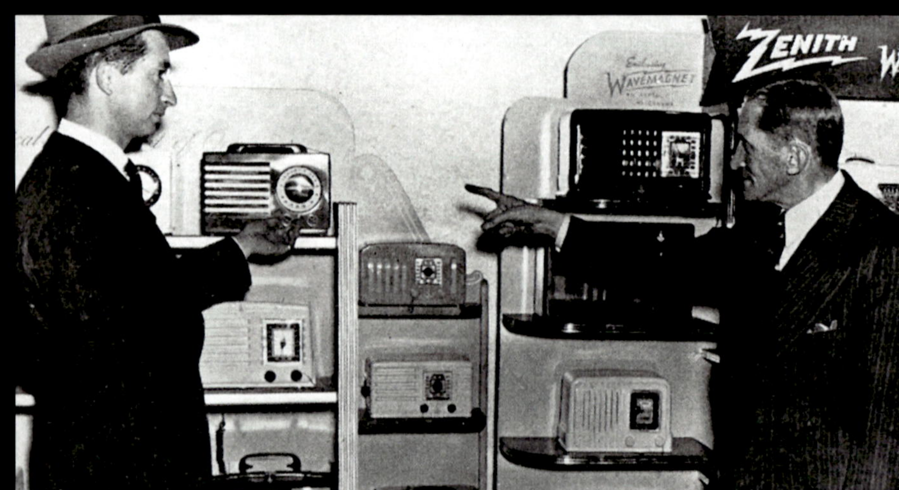

EMERSON "Aristocrat"
USA
Designer: Norman Bel Geddes

1941

1941 MARCONI
Canada

EGM 1941
Mexico

1941 ZENITH 4B515
USA

FADA Bullet 189 "All American"
USA

1941

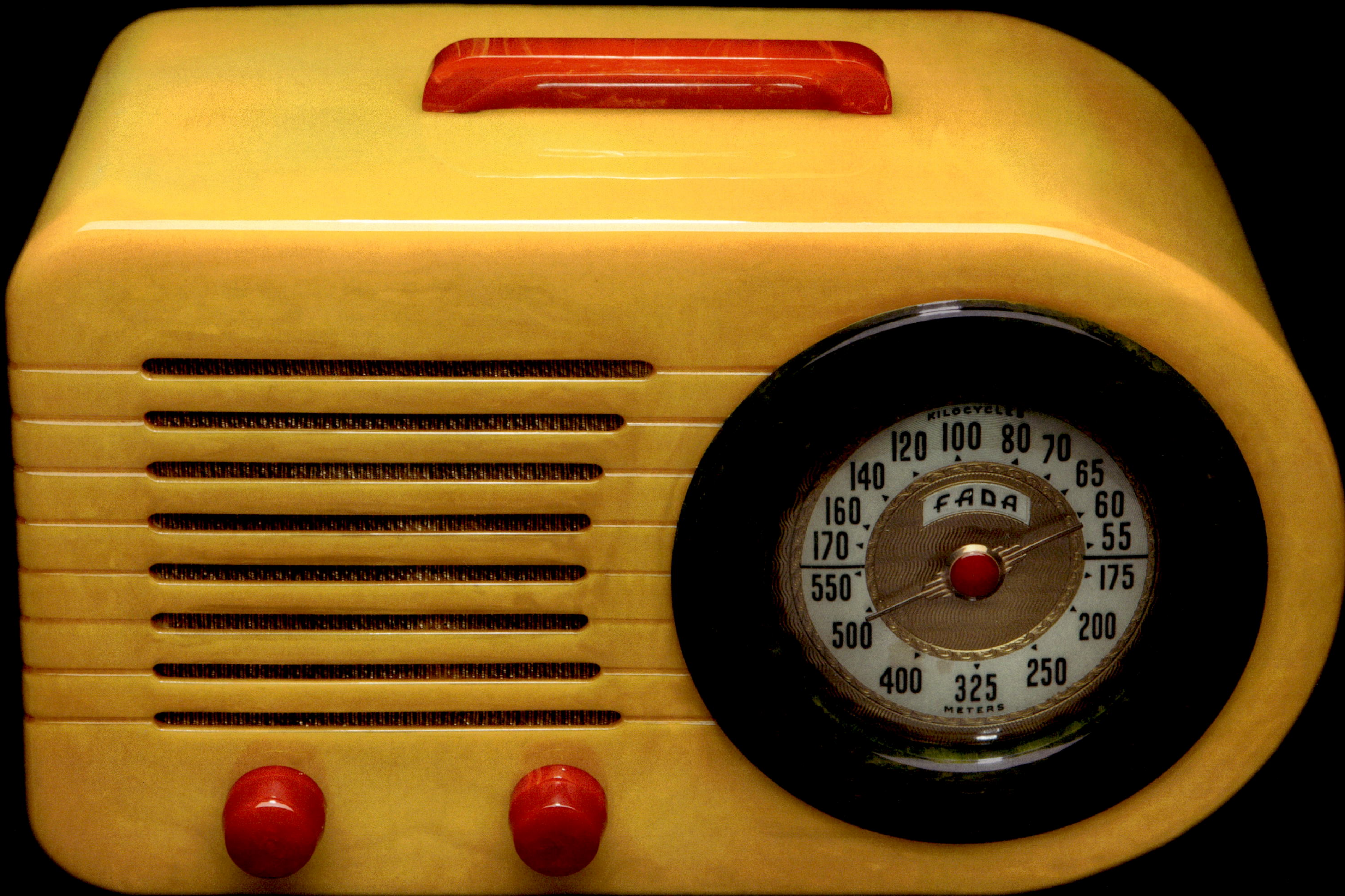

FADA Radio Presents

Gem-like Beauty

IN NEW 1941

Catalin Radios

A Symphony in Color

REPRINT

THE RADIO OF TOMORROW •• TODAY

FADA Radio

"BEAUTIFUL!... EXQUISITE!... HANDSOME!... "GORGEOUS!... GEM-LIKE!... and HARMONIZING!..." are just a few of the exclamations made by those who have seen these

New 1941 Fada CATALIN MODELS

MODEL 115A—Alabaster Cabinet with translucent illuminated DIAL. Same cabinet for Model 116A, but dial has different numerals for short wave.

MODEL 115XA—Onyx Cabinet with Alabaster appointments and translucent illuminated DIAL. Same cabinet for Model 116XA, but dial has different numerals for short wave.

MODEL 115MA—Ruby-like Maroon Cabinet with Alabaster Appointments and translucent illuminated DIAL. Same cabinet for Model 116MA, but dial has different numerals for short wave.

MODEL 115BA—Blue Lapis Lazuli Cabinet with Alabaster Appointments. Translucent illuminated DIAL. Same cabinet for Model 116BA, but dial has different numerals for short wave.

THE RADIO OF TOMORROW •• TODAY

FADA Radio

115 SERIES CATALIN GROUP

5 Tube AC-DC Superheterodynes with FA-DA-SCOPE LOOP ANTENNA & DROP HANDLE

Your choice of FIVE color combinations in *beautiful gem-like* CATALIN CABINETS that resemble many precious stones such as onyx, alabaster, ruby and lapis lazuli.

ALL MODELS IN THIS GROUP **$19.95**

116 SERIES (Short Wave) CATALIN GROUP

BROADCAST and FOREIGN RECEPTION

IN THE SAME BEAUTIFUL CATALIN CABINET STYLES AS 115 SERIES. Provides additional reception from International Short Wave Stations.

ALL MODELS IN THIS GROUP **$22.95**

See Reverse Side for Complete Description and Specifications

FADA 115/1000 (Bullet) 1941/1945
USA

c1941 STROMBERG CARLSON prototype
USA

SONORA KM "Coronet" 1941
USA

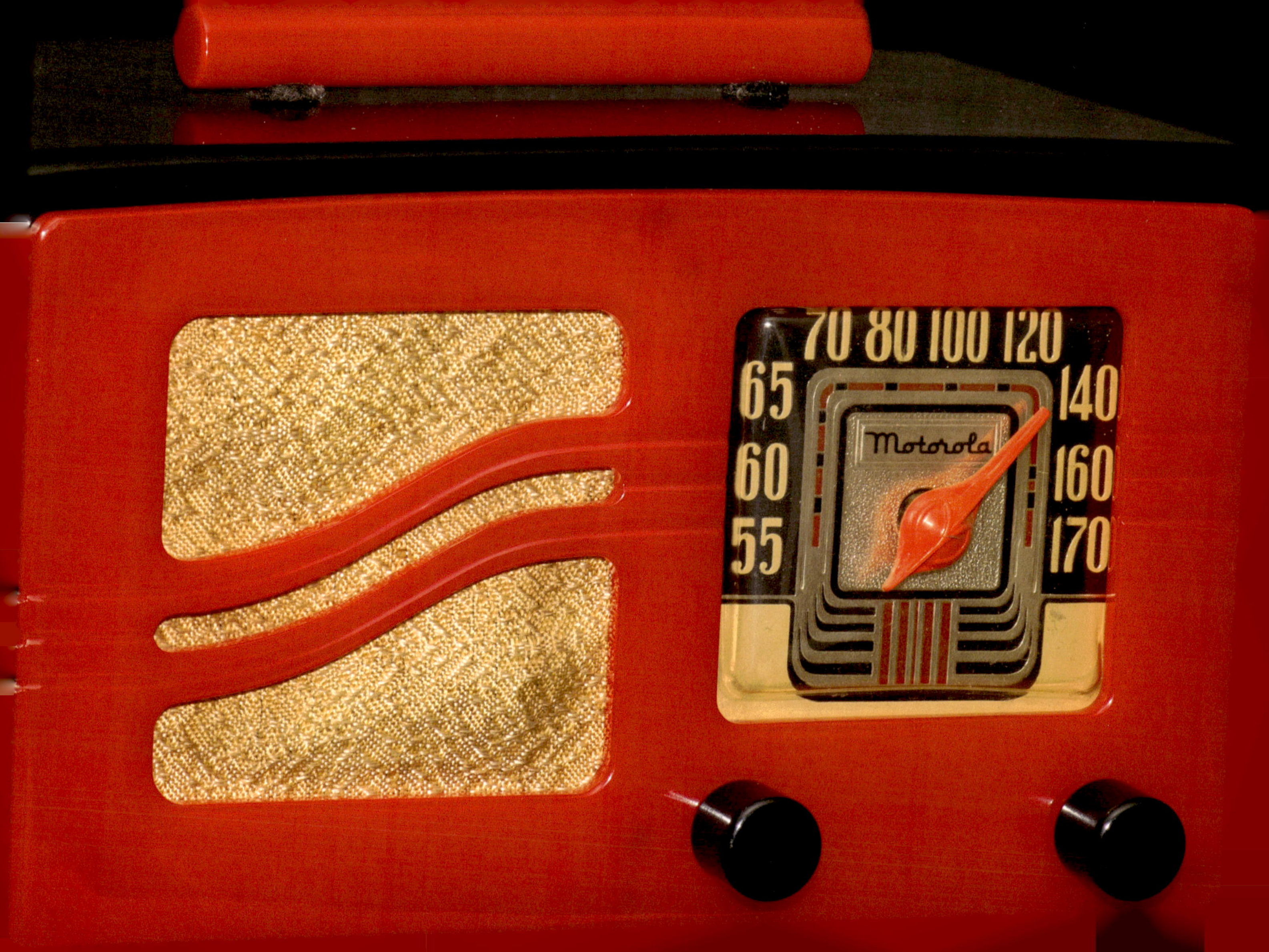
70 80 100 120
65
60
55
140
160
170
Motorola

MOTOROLA 51X (S-Grill) 1941
USA

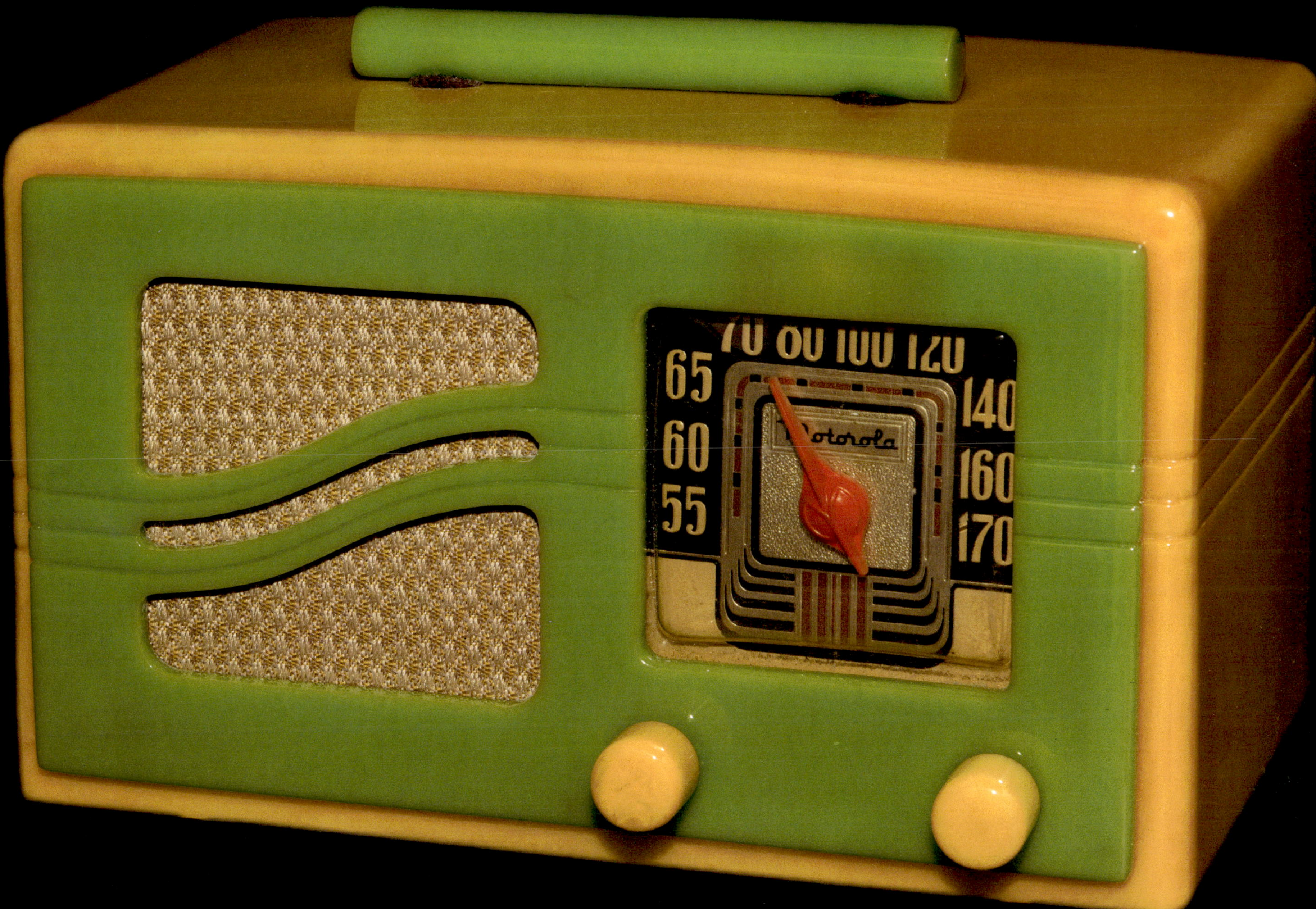

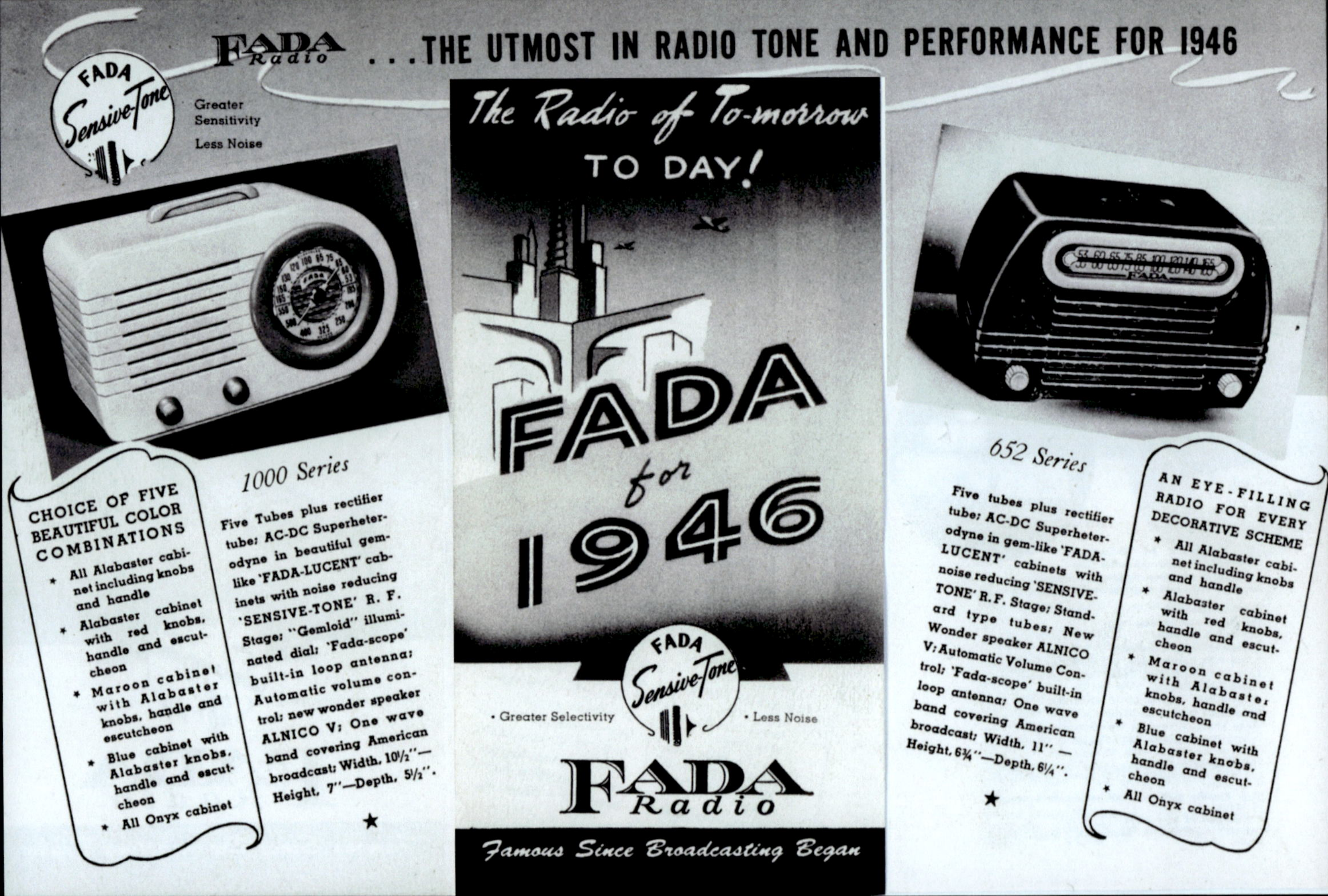

Fada RADIOS

Model 1000 is one of series available in several color combinations. 6-tube ac-dc superhet. Features "Sensive-Tone". OPA price $35.50.

Model 652 is one of series in various color combinations. 6-tube ac-dc circuit gives 8-tube reception. Features "Sensive-Tone" for greater sensitivity. OPA price $36.50. Fada Radio & Electric Co., Inc., 3020 Thomson Ave., L. I. C., New York—RADIO & Television RETAILING

FADA 652 (Temple)
USA

1945

1945 GAROD 6AU1 "Commander"
USA

EKCO A22 1945
UK
Designer: Wells Coates

LIFE TESTED*

for Reception Perfection!

Every Sentinel radio—table model or console, portable or radio-phonograph—is Life Tested *before* being assembled into the finished product. There is constant testing and re-testing—proving and improving—to assure the Reception-Perfection—the *Studio Tone In Your Home*—for which Sentinels have become so widely celebrated!

Unequalled in tonal clarity and quality, Sentinels are likewise stand-outs in styling—with models for every room in every home, for every purse and purpose. Modern or classical, in subdued or gaily colored plastics, in richly grained woods—there is a Sentinel to match the mood of your decorative scheme . . . to delight your eye, to compliment your ear—and to flatter your sense of value!

SENTINEL RADIO CORPORATION
EVANSTON, ILLINOIS

*Life Testing of every part *in advance* ascertains that every Sentinel radio set will prove superior to any stress or strain, any wear or tear to which they will ever be subjected short of absolute abuse. This testing and re-testing is applied, not alone to the complete radio but to all electrical and mechanical parts, to cabinets as well—even to tuning knobs!

for Studio Tone in Your Home!

Sentinel *Radio*

SENTINEL 284NI
USA

1945

1945 STEWART WARNER 62T36
USA

NAMCO
USA

1946

1946 PHILCO 46-420 (Hippo)
USA

ČSR 197.7 Frankfurt ČSR 243.5
O.C. 188
ČSR
Hörby
Zagreb
Bratislava
Wrocław
Leipzig I
Graz-Dobl
Berlin
Brno
London
Bucureşti I
Roma I
Sofia I
Rennes I
Stockholm
Hilversum I
ČSR 428
Beograd I
Marseille I
Charkov
Praha
Bruxelles I
Brasov
Strasbourg I
Moskva I
Ankara
Motala
Droitwich I
Oslo
Warszawa I
Kalundbg.
Lathi
D. Sender
ČSR 1102.9
Minsk
Beromün.
Budapest I
Riga
Helsinki
Lyon I
Wien I
TESLA
SACKVILLE
NEW YORK
MOTALA
MOSKVA
SCHENECTADY
LONDON
SHEPPARTON
MONTE CARLO
PRAHA
ALLOUIS
MOSKVA
WIEN
LONDON
187-571m
1000-2000m

ÉLÉGANCE ET
QUALITÉ DANS
L'INÉDIT
le Super
BOUM
Toute la technique
de ses aînés
C'est une création "RADIALVA"
Radialva
1, RUE J.J.ROUSSEAU • ASNIÈRES (SEINE) GRÉ. 33-34

RADIALVA "Super Boum" and speaker 1946
France

1946 EMERSON 578
USA
Designer: Charles & Ray Eames

RCA 66X8 (Tuna Boat) 1946
USA

Why

there are so many types of BAKELITE Plastics

THE SWIFT PACE of the Plastics Industry, with new developments and discoveries crowding one upon the other, has raised in the minds of businessmen everywhere the question, "Why Are There So Many Types of BAKELITE Plastics?"

The answer goes back to the turn of the century when the introduction of the *Phenolics* gave the infant industry its greatest impetus. Uses by the thousands were found for these versatile materials, which, even then, were supplied in several forms, suitable for molding, laminating, casting, coating, and bonding.

Through the years, manufacturers' requirements became increasingly more exacting and other types of plastics were created to meet their specialized needs. This is demonstrated by reviewing only one group of plastics—the molding materials.

The *Phenolics* were produced mainly in opaque blacks, browns, and darker colors, because of the nature of the basic resins. The call for brilliant colors and translucent effects was met by the *Ureas.* Next, color had to be combined with a high degree of toughness and impact resistance, and the *Cellulose-Acetates* were introduced. Then, came demands for transparent, translucent and opaque materials with improved physical, chemical, and electrical properties, and the *Polystyrenes* were developed.

BAKELITE materials are produced in a variety of forms for the convenience of manufacturers using different methods of fabrication. The illustration features four of these forms—cast resins, molding materials, varnishes for laminated plastics, and synthetic resins for surface coatings. In addition, Bakelite Plastics Headquarters produces impregnating and calendering materials, cements, adhesives, and bonding resins.

For further information on these materials, write for a copy of Booklet 13P, "NEW PATHS TO PROFITS," the businessman's guide to modern plastics.

BAKELITE CORPORATION
Unit of Union Carbide and Carbon Corporation
UCC
30 EAST 42ND STREET, NEW YORK

BAKELITE

TRADE MARKS
The word "Bakelite" and the identifying products Symbol are registered trade-marks of Bakelite Corporation

PLASTICS HEADQUARTERS

The ever-broadening diversification of materials offered by Bakelite Corporation is now supplemented by the "Vinylite" plastics developed and produced by Carbide and Carbon Chemicals Corporation. They form an entirely new and useful group of materials for modern living. Both of these companies are units of Union Carbide and Carbon Corporation.

GENERAL TELEVISION
USA

1946

CYARTS B
USA

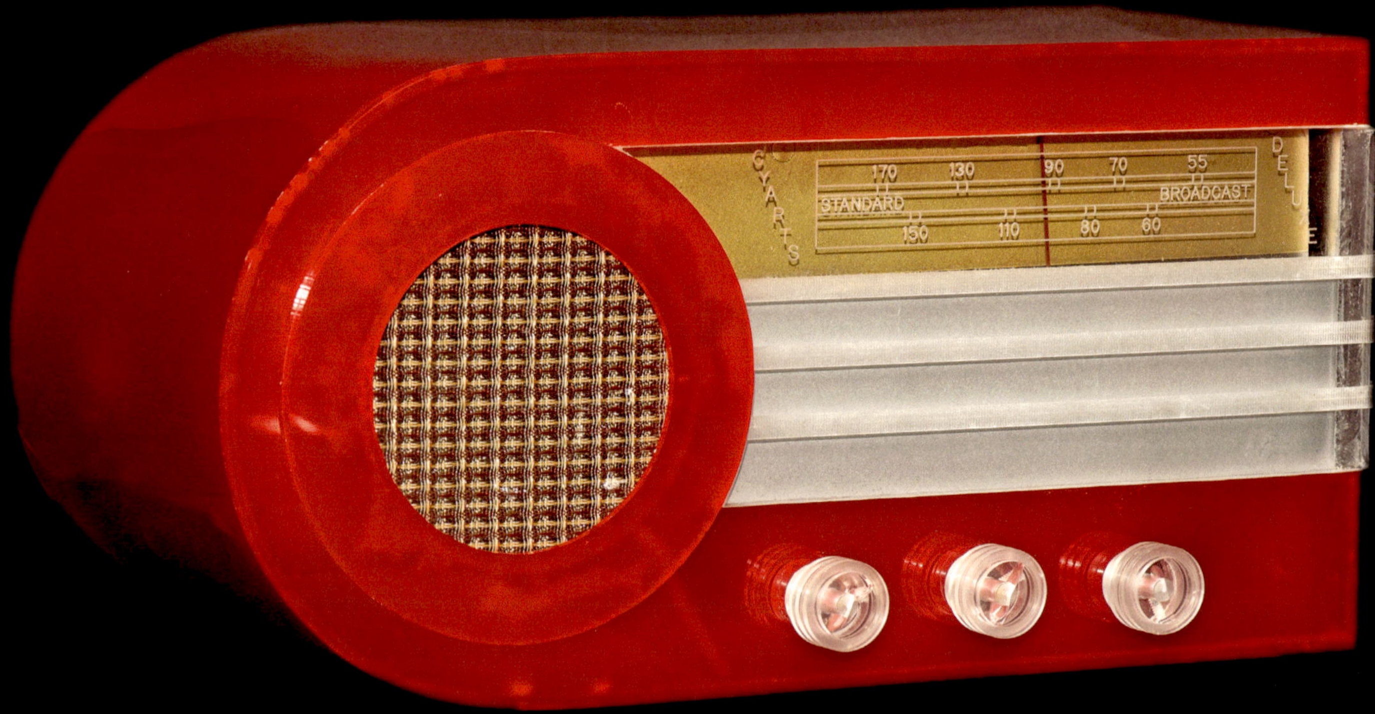

FADA 700 (Cloud)
USA

1946

1946 BELMONT 6D-111
USA

AIR KING A-600 "Duchess"
USA

1946

1946 BUSH DAC90
UK

STC "Bantam"
Australia

1946

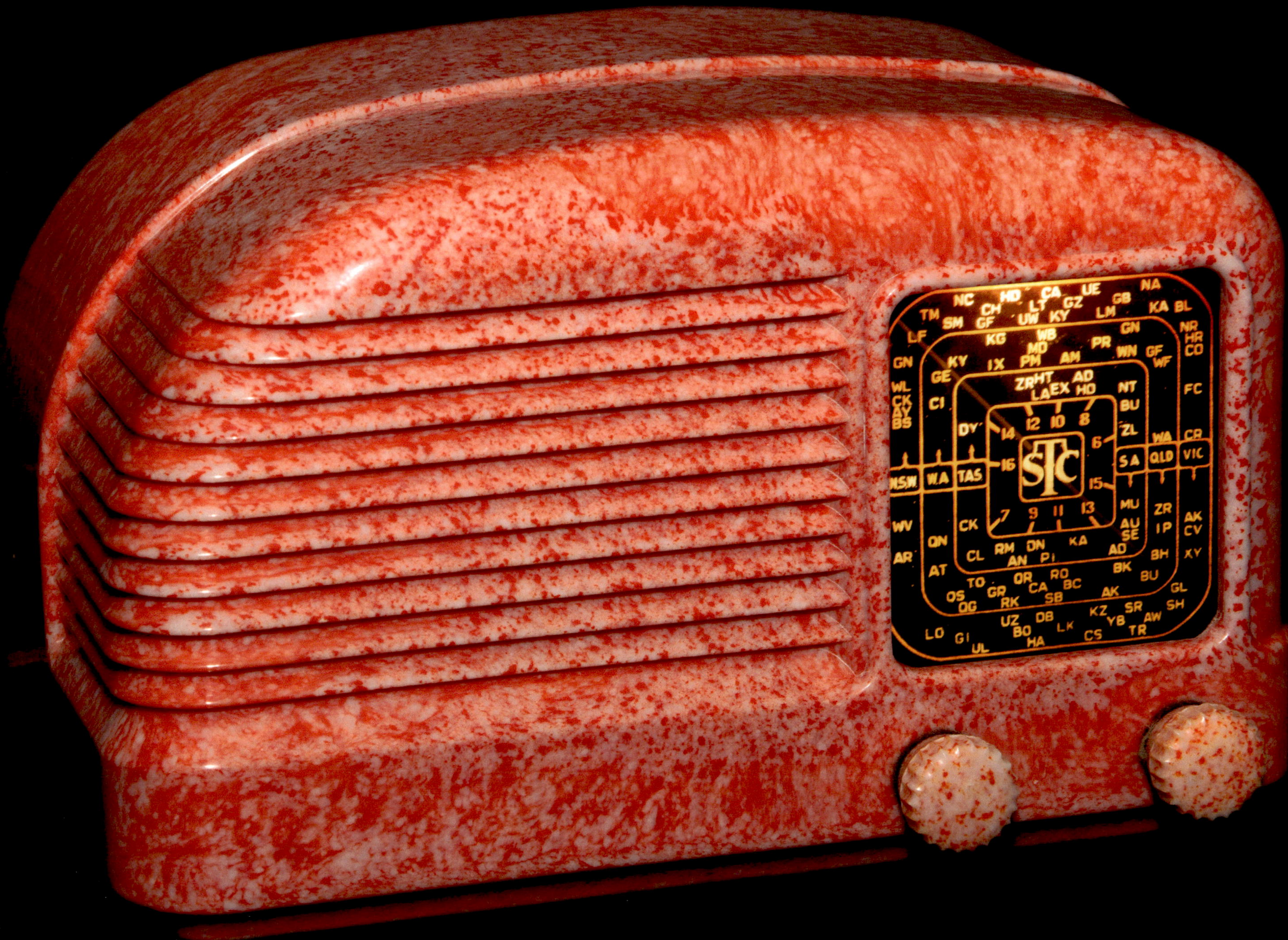

1946 BENDIX 526C
USA

ARTES AR3 1947
Spain

1947 MULLARD "Meteor" 600
Australia

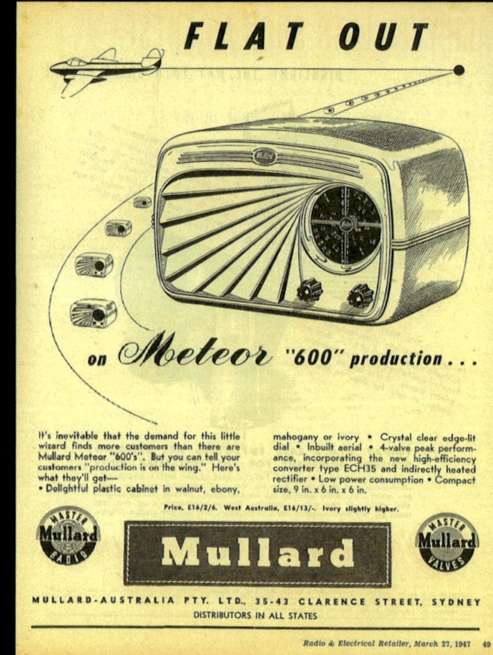
FLAT OUT

on *Meteor* "600" production . . .

It's inevitable that the demand for this little wizard finds more customers than there are Mullard Meteor "600's". But you can tell your customers "production is on the wing." Here's what they'll get—

• Delightful plastic cabinet in walnut, ebony, mahogany or ivory • Crystal clear edge-lit dial • Inbuilt aerial • 4-valve peak performance, incorporating the new high-efficiency converter type ECH35 and indirectly heated rectifier • Low power consumption • Compact size, 9 in. x 6 in. x 6 in.

Price, £16/2/6. West Australia, £16/13/-. Ivory slightly higher.

MASTER Mullard RADIO

Mullard

MASTER Mullard VALVES

MULLARD-AUSTRALIA PTY. LTD., 35-43 CLARENCE STREET, SYDNEY
DISTRIBUTORS IN ALL STATES

Radio & Electrical Retailer, March 27, 1947 49

SONORA WAU-243
USA/France

1947

1947

EMERSON 511 "Moderne"
USA
Designer: Raymond Loewy

KOLSTER BRANDES BM20 (Dalek)
UK
Designer: Lawrence Griffen

1947

1947

STEWART WARNER "Air Pal"

USA

THE SATURDAY EVENING POST

Smaller than a phone ...performs like a console radio!

ONLY Stewart-Warner GIVES YOU *Strobo-Sonic Tone*

So faithful you can understand each word, distinguish every instrument, hear music in its true dimensions, rich and picture-clear. Only Stewart-Warner has it.

FM

Stewart-Warner

Frequency Modulation brings you the beautiful harmonic overtones, the real-life quality of every instrument and voice, against a background of velvet silence. No static.

Balladeer

FM-AM TABLE RADIO

Genuine FM reception in addifion to all standard broadcasts. AC-DC. For living-room, kitchen, bedroom . . . smart new table styles in both wood (walnut or desert tan) and plastic (walnut or ivory).

Musidora

FM-AM COMBINATION CONSOLE

A master instrument for FM, standard radio, and records. Intermix record changer, 2 built-in aerials, 2 compartments. Wide-range AC reproduction. In fine mahogany.

Air Pal *plays anywhere you plug it in*

Engineered like a watch, Air Pal is a full-power AC-DC radio that weighs only 3¼ lbs. Yet it has vernier planetary tuning, accurate slide-rule dial, full AM band, built-in Radair Antenna, genuine dynamic speaker, 4 tubes *plus* rectifier. Full-power performance—*not a toy with tinny tone.* Fits in a handbag. Plays in trains, hotels, offices, at home.

Ebony, walnut, or ivory plastic.

Videorama Television

with FM Sound plus AM Broadcast Band—Direct-View Pictures

Clear, steady pictures under living-room lights. Big 10-in. cathode-ray tube. All 13 channels. In modern or Georgian cabinets.

A MODEL FOR EVERY PURPOSE...A RADIO FOR EVERY ROOM

Stewart-Warner

AM *Radios* FM

Radio-phonographs · Television

The Consolette

New Combination Radio-Phonograph

Full-voiced console-quality tone...smooth AC operation for full-cycle tonal range. Plays up to 12 records automatically. Four quickly detachable legs make this model a smart table set.

CHICAGO 14 ILLINOIS

ASTOR "Baby" (Football) 1948
Australia

1947 SONORA "Excellence" 301
France

1948

ARVIN 442
USA

TEMPLE Superhet
USA

PYE

Good Radio in Miniature

MODEL M78F

Your "Personal" receiver, battery operated

OPERATING INSTRUCTIONS AND BRIEF DESCRIPTION

PYE M78F 1948
UK

1948 ASTOR JN (Harbour Bridge)

Australia

PHILCO "Transitone" 49-501 (Boomerang) 1949
USA

1949 ZODIAC "Moskito"
France

1950s

USA
Sterling Deluxe
EMERSON
FRANCE
NEW ZEALAND
Pacemaker
AUSTRALIA
SPAIN
UK
CZECHOSLOVAKIA
RUSSIA

1950

TELEFUNKEN 1065U "Bahia"
Spain

ESPEY 30 "Roundabout" 1950
USA

1950

ARVIN 740
USA

ARVIN 540
USA

SILVERTONE
USA

ARVIN 542
USA

HEALING "Moderne" (Scales) 1950
Australia

1950 SONORA "Sonorette" 50
France

Sterling
Deluxe

1950 KOLSTER BRANDES FB10 (Toaster)
UK

CROSLEY (Bullseye)
USA

1950

c1951 PACEMAKER
New Zealand

TESLA "Talisman" 306-U
Czechoslovakia

1951

1954 EMERSON 744B
USA

ZVEZDA "Red Star"
Russia

1954

1954 SPARTON "Easy-Goer"
USA

OCEANIC "Surcouf"
France

1956

BIBLIOGRAPHY & RECOMMENDED READING

Albrecht, Donald (Editor). *Norman Bel Geddes Designs America*. New York, NY: Abrams, 2012.

American Industrial Designers. Richardson, TX: Hephaestus Books.

Benton, Charlotte, Tim Benton, and Ghislaine Wood. *Art Deco 1910–1939*. London, UK: V&A Publishing, 2003.

Cohn, Laura. *The Door To A Secret Room. A Portrait of Wells Coates*. Aldershot, UK: Scolar Press, 1999.

Collins, Philip. *Radios—The Golden Age*. New York, NY: Black Dog & Leventhal Publishers, 1987.

Collins, Philip. *Radios Redux: Listening in Style*. San Francisco, CA: Chronicle Books, 1991.

Fenichell, Stephen. *Plastic—The Making of a Synthetic Century*. New York, NY: HarperCollins, 1996.

Fiell, Charlotte and Peter Fiell. *Plastic Dreams: Synthetic Visions in Design*. London, UK: Fiell Publishing, 2009.

Ganz, Cheryl. *The 1933 Chicago World's Fair*. Chicago, IL: University of Illinois Press, 2012.

Hanks, David A. and Anne Hoy. *American Streamlined Design. The World of Tomorrow*. Paris, France: Editions Flammarion, 2005.

Hawes, Robert. *Radio Art*. London, UK: Green Wood Publishing, 1991.

Hawes, Robert and Gad Sassower. *Bakelite Radios*. Edison, New Jersey: Chartwell Books, 1996.

Hiller, Bevis and Stephen Escritt. *Art Deco Style*. London, UK: Phaidon Press, 1997.

Innes, Christopher. *Designing Modern America: Broadway to Main Street*. New Haven, CT: Yale University Press, 2005.

Johnson, J. Stewart. *American Modern 1925–1940 Design For a New Age*. New York, NY: Harry N. Abrams Publishing, 2000.

Katz, Sylvia. *Early Plastics*. Buckinghamshire, UK: Shire Publications, Nd.

Katz, Sylvia. *Classic Plastics: From Bakelite to High-Tech*. London, UK: Thames and Hudson, 1985.

Loewy, Raymond. *Industrial Design*. New York, NY: Overlook Duckworth, 1979.

Mabbs, Merrill. *Classic Radio Gallery: Classic Plastics Vol. 1*. USA: Bozbuilt Productions, 2011.

Meikle, Jeffrey L. *American Plastic: A Cultural History*. New Brunswick, New Jersey: Rutgers University Press, 1995.

Meikle, Jeffrey L. *Twentieth Century Limited: Industrial Design In America, 1925–1939*. Philadelphia, PA: Temple University Press, 2001.

Mossman, Susan. *Early Plastics*. London, UK: Leicester University Press, 1997.

Perree, Rob. *Bakelite: The Material of a Thousand Uses*. Amsterdam, Holland: Cadre/Snoeck-Ducaju and Zoon NV, 1996.

Sheridan, Peter and Ritchie Singer. *Radio Days: Australian Bakelite Radios*. Sydney, NSW: Bakelite, 2008.

Sideli, John. *Classic Plastic Radios of the 1930s and 1940s*. New York, NY: E. P. Dutton, 1990.

Slusser, John. *Collector's Guide to Antique Radios 7th Ed.* Paducah, Kentucky: Collector Books, 2008.

Wilk, Christopher (Editor). *Modernism: Designing a New World*. London, UK: V&A Publishing, 2006.

Wilson, Guy, Dianne Pilgrim, and Dickran Tashijian. *The Machine Age in America: 1918–1941*. New York, NY: Brooklyn Museum and Harry N. Abrams Publishing, 1986.

GALLERY INDEX

GENERAL INDEX

Sentinel
STANDARD BROADCAST
55 60 70 80 90 110 130 150 170
MADE IN U.S.A.
FADA
KILOCYCLES
METERS
Sparton
ARVIN
STEWART-WARNER
Pacemaker
STEWART
BROADCAST
SHORTWAVE
55 60 70 80
The Fisk Radiola
ATLAS
MEGACYCLES
KILOCYCLES
SYDNEY
MELBOURNE
BRISBANE
ADELAIDE
PERTH
HOBART
VICTORIA
QUEENSLAND
STOCKHOLM
PARIS P.T.T.
BEOGRAD
CAIRO
MOSKWA
LISBOA
TOKIO
DAVENTRY
BUDAPEST
LONDON
BERLIN
MILANO 1
TOULOUSE
MÜNCHEN
ROMA 1
SOTTENS
NEW YORK
ZEESEN
VATICANO
NORMAL
KILOCICLOS
METROS